A GPS User Manual

Working With Garmin Receivers

By

Dale DePriest

The author believes the data in this manual to be correct at the time of writing, but there is no warrantee expressed or implied as to the fitness or accuracy of this document for any purpose. Neither the author nor the publisher shall have any liability or responsibility for the use or suitability of any of the material or methods described in this book, nor for the products described.

Garmin is a registered trademark of Garmin LTD. Garmin product names are registered trademarks of Garmin. Some Garmin copyrighted images were used in the generation of the cover and other figures in this document. These images are copyright Garmin LTD. or one of its subsidaries and are used with permission.

1stBooks - rev. 05/10/03

Preface

This is a user manual for Garmin receivers. It covers all Garmin handheld receivers from the older multiplex units to the latest and greatest full-featured units, which includes the latest Garmin units at the time of printing, the etrex, the emap, and the 76 family. This manual does not cover the street pilot series since they are not handhelds, specific marine units, or dedicated aviation units (except the III Pilot as an aviation example), but you should find the general information to be applicable to any Garmin unit. And, a chapter is included at the back that describes the general features of these units and some others. Since I don't own every model of Garmin receiver, and since some simplification of details is necessary in a manual with this much scope, there could be small discrepancies between this presentation and the way the product really works. In addition, the products change over time as Garmin modifies the GPS models behaviors with new firmware updates. Therefore, I can make no guarantee that the descriptions in this manual exactly match the product you may have purchased, but I have tried to make it as accurate as possible. The information in this manual is based on Garmin documentation, my own experiences, and data collected from the GPS newsgroup called sci.geo.satellite-nav. This is a Usenet newsgroup and is a great source of information about GPS units and GPS topics in general. This manual is designed to complement the Garmin supplied user manual but is intended to be complete enough that it could substitute for that manual if you don't have one. You could also download the appropriate manuals from the Garmin web site, http://www.garmin.com.

The scope of this manual is significantly beyond the scope of a typical user manual in that it describes how a Garmin GPS works and how to use it to do what you want to do. Generally most manuals only attempt to show what the commands do without much regard to how they are intended to be used or how they actually work. In this manual there is only one chapter that describes how the commands work while most of the rest of the manual shows you how to use a Garmin GPS to do things you might want to do. Even the user interface chapter generalizes the command descriptions by showing the philosophy of the interface design. Once the philosophy is understood it is hoped that the user will be able to figure out how to get the solution for their needs even if they forget the exact command sequences. The manual organization and presentation style would also be suitable to use as a textbook in a class devoted to using GPS receivers. In this case the style permits each student to have a different model unit and still be able to follow the course without having to describe each command sequence for the various models over and over again. The theory of GPS operation is also covered since this is necessary to understand if you want to know the limitations of this technology.

Most readers are likely to already own their Garmin receivers but just in case a reader needs some help in selecting a GPS receiver there is a chapter devoted to this subject near the end of the book.

I would like to offer my special thanks to the sci.geo.satellite-nav newsgroup whose members have provided me with the GPS education I needed to write this manual. Any one wanting further information should begin their research at the web site http://gpsinformation.net.

Note that information provided in this manual is based on my own knowledge and, in some cases, conjecture. Garmin considers much of its internal algorithms to be proprietary so a user must look at the results and try and guess as to how they were computed. Garmin has every right to maintain this position, but it means that there is some subtle performance information that is undocumented by them. I have attempted to provide one reasonable explanation for these kinds of behaviors but it may not be the correct one. It should, at least, help you in understanding how it could work!

Some drawings in this manual were captured in g7to, I drew some, and vendors supplied some. The information in the WAAS section is derived from the FAA technical specification, available from the FAA web site, other sources, and personal observations. The US Coast Guard web site provided some of the information in the beacon section. The G-12 does not support screen capture directly but James Associates supplied the screen shots.

A special thanks to Marvin Thorman for the idea behind the technique presented in the route chapter on adding points from the map screen, and to Jerry Wahl for the idea behind the user defined grid discussion and example. Thanks to Joe Mehaffey and Jack Yeazel for their web site full of GPS data and the privilege to add my stuff.

Table of Contents

List of Figures

Chapter 1

Introduction

"Working with Garmin receivers" is a practical operation manual for the entire line of Garmin handheld receivers. While Garmin has many features that are the same across their entire product line they also have many differences. In an attempt to isolate the major user interface differences to one section, this manual uses fairly generic step by step instructions throughout the document. Please read the section on user interface early in your reading. This will help you understand the philosophy of the interface design and will allow you to understand generic instructions and relate them to the exact key sequences that you will need to use.

This manual covers both older 'multiplex receivers' and the newer 12 channel parallel units. Older units are primarily the G-45 family of products. The 45 family consists of the 45 and its successor the 45XL as well as its brothers the G-40 and the baby of the family the G-38. For automobile use many chose the G-II as their preferred unit. At one time these were the premier handhelds but technology has made them obsolete but still useful for a great many people.

Figure 1 - G-12 and G-III+

The newer parallel units fall into several families. The G-12 family looks similar to the older multiplex units and includes the G-12 at the bottom (shown on the left in figure 1) and the G-12XL as the vanguard product. The G-48 rounds out the family at the top as a replacement for the older G-45XL. To maintain the automotive market the G-II was replaced with the G-II+, still a favorite for users wanting a unit that can switch itself from automotive (horizontal orientation) to handheld (vertical orientation). A new family was introduced with the G-III (shown on the right). It uses the case design of the II+ and added internal maps to the suite of other information available for the user. The G-III family includes the G-III, the G-III+ with uploadable detailed street maps, and the G-III Pilot, with a Jeppesen database and special features for flying use. The G-III family has a different interface from the G-12 family in that it is much more menu oriented.

In addition to these families there are several individual products that are covered in the manual. The 12CX is a new handheld with color display, 1000 waypoints, a 2000 point track log, audible alarms and dedicated zoom keys. Otherwise it is similar to the 12XL. The 12MAP uses the 12CX hardware design with its audible alarm and patch antenna. It is then loaded with III+ software and high-resolution gray scale screen so it has built in maps, customizable display features and III family user interface.

Figure 2 emap, etrex, 76

Garmin introduced a pluggable memory cartridge in its Street Pilot vehicle units and later developed the emap as a portable unit that can use the same cartridges (shown above to the left). It differs from other Garmin units in many ways and is focused on providing an electronic map display with GPS functionality primarily for vehicle use although many like the fact that it is thin and goes well in a pocket.

Garmin took a new tack in the etrex. This began as a small entry level hiking unit weighing only 5.3 oz (shown above in the middle). Its success led to the introduction of a whole new family of smaller handhelds including the Summit, Garmin's first unit with an integrated fluxgate compass and barometric altimeter. Then came the Venture, a full-featured replacement for the G-12 series for hiking use; and two new mapping units, the Legend and the Vista. The Vista combines a mapping unit with an integrated compass and altimeter. To provide for marine users Garmin has introduced the 76 (shown above on the right), the Map76, and the Map76S family of handhelds with specific marine features. The 76S has a compass and altimeter built in. These units are also favorites for other users that prefer a larger display as compared to the etrex.

Thus Garmin handhelds have progressed from the original multiplex receivers to the second generation of receivers featuring 12 channel receivers to the latest even lighter units that all run off of 3 Volts power while supporting the same or improved functions of their 6 volt predecessors.

This manual does not explicitly cover non-handhelds like the StreetPilot family or the new aviation/automotive unit, the 295. Nor does it cover the dedicated marine units. There is a chapter briefly highlighting these units so that these users can determine which areas of the manual are likely to apply to their models. The StreetPilot units have a distinct user interface and different capabilities that primarily feature detailed moving map support with secondary support for traditional GPS functions. They are most like the emap unit, which is covered in this manual. Users of the aviation or larger marine products will find that these products are similar in many ways to the handheld units and thus these user may still find this manual useful in describing features and capabilities as generally applied to the Garmin philosophy and theory of GPS design.

Getting Started

The first step in working with your new Garmin is getting power for the unit. Most of the time this means installing a set of 4 AA batteries (2 AA's on the latest generation units). To install the batteries look on the bottom of the unit and turn the D ring 90 degrees counterclockwise. (The emap door just slides off.) This will release the door and permit the installation of the batteries. Notice that there is a + and a - indicated near the battery tubes. Make sure the new batteries a placed in the unit with the + side of the battery showing at the + end and the - side of the battery showing on the minus end. Once the batteries are installed you can close the case using the D ring to lock it down. All Garmin handhelds also support an external interface connector that can also be used to power the unit. This use is covered in a later chapter.

Now turn the unit on by holding the red button down long enough for the opening screen to appear on the display. (The power button is on the side on etrex and emap.) The unit will perform some internal tests and start acquiring satellites so that it can compute a position. Mapping units will normally require you to acknowledge a disclaimer screen by pressing the enter key. If you wait long enough this screen will time out. If this is the first time you have used the unit you will normally have to provide it with a general idea of your current location. This is an optional step, and is not available on the etrex, since if you are willing to wait a little longer to find your first position you can skip this. This can be done on some units by scrolling through a list of states and countries, using the arrow keys or pad, or on units with maps you can scroll on the map itself to select your approximate location. (For more information on obtaining a fix see the chapter on this subject.) Once the unit has acquired 3 satellites it will compute a fix and switch from the opening satellite status screen to the position screen. (The emap does not switch screens nor does it have a page key.) You can now use the unit to read your current location and other data. Pressing the page key multiple times will switch you through most of the available screen displays.

To turn the unit off, press and hold the red button (the same button you used to turn it on) for 3 seconds until the unit turns off. While the unit is on, pressing this button will toggle the backlight on and off.

(Emap has a separate button for this.) On some units you can press this button multiple times to increase the lamp brightness. There is a small icon that looks like a light bulb on the opening status screen (The etrex and emap have this on the global menu screen.) that indicates when the lamp is on. Generally this lamp is set to timeout after a period of time. If it goes out, depressing any key will turn it back on. You will have to depress the key a second time to perform whatever function it was supposed to do.

While any AA cells can be used in your unit it will work best with AA alkaline batteries. Four fresh batteries will generally produce slightly more than 6 volts of output and register full on the battery gauge that appears on the status screen. (Two batteries for 3 volts on the etrex, emap, and 76) The empty mark on the gauge corresponds to 4 volts (2 volts on the etrex, emap, and 76) from the batteries. The unit will shut down when the battery voltage drops below about 3.8 volts on a G-III family, or about 3.7 volts on a G-12 family or one of the older multiplex units. Battery life varies from one unit to another with newer units getting the best battery life. On older units the battery life can be as low as 8 hours while on the latest units using a battery save feature can reach a claimed life of 35 hours from a 4-cell unit and nearly 20 hours on a 2-cell unit. Actual battery life varies a lot depending on conditions.

Many users prefer to use rechargeable batteries in their units. While they need to be replaced more often the total life of the batteries is much longer. The low battery warning level works reasonably well for all battery types but the full charge on ni-cad and NiMH batteries will show only 3/4 full. This is normal behavior, but the G-III family and some other models including some models of the etrex family have a special setting for ni-cad batteries to give a better battery display if you want to use it. NiMH is also a setting on a few models. Some users have reported longer life using the regular setting for NiMH with some models. For best battery performance you should replace rechargeables as soon as you receive the low warning indication. You will have to recharge any batteries external to the unit as there is no provision for recharging them while inserted in the case. In cold climates regular lithium batteries may offer the best performance. There are no rechargeable lithium batteries that will work in these units. Some models have a special battery setting for Lithium cells as well.

Battery manufacturers are not very consistent in the diameter of the AA batteries. In particular the batteries with the built-in gauges tend to be slightly larger in diameter than the normal batteries. This has been known to cause problems of premature shutdowns in some units in high vibration environments, such as on motorcycles or off road vehicles. The best solution, if you experience this problem, is to add a ring of tape to the outside two batteries so that they are fairly snug in the tube. Don't add too much or you may have problems removing the batteries. Garmin retrofitted some G-II+ units and G-III units with a capacitor to minimize the impact of this problem but you may still experience it in a few cases. Using external power is also a good way to prevent this problem. For more information on batteries look in the User Interface chapter.

What you can do with a GPS and why you would want one

Here is a list for drivers and RV owners.

1. They are a great deal of fun to use.
2. They are useful if you get lost and you have a map that gives UTM or lat/lon coordinates.
3. You can actually follow yourself on a map in real time display which allows planning turns, stops, etc if your unit has output capability.
4. You can answer the question "Are we there yet?"
5. Did I mention they are fun?
6. You can find where you left your car.
7. Lets you tell the AAA where to send the tow truck.
8. You can set waypoints and see directions to get there.
9. You can record the exact path you took to get somewhere and play it back later for someone else to see.
10. You can prove how fast you were going when stopped by the cops.
11. You can use it as a compass (if you're moving).
12. When camping, you can use it to find your camp when you go on a hike or to town.
13. You can use it to estimate arrival times.

14. Did I mention they are fun?
15. You can calibrate your speedometer.
16. You can use it as a speedometer if yours is missing or broken.
17. You can use it as a trip meter.
18. With the right database you can find restaurants and other things.
19. You can plan a trip ahead of time and the unit will even tell you where to turn.
20. You can use the info from a GPS to program your dbs satellite receiver.
21. If it is connected to a cell phone you can give the police your cell phone number and they can track your stolen vehicle (some extra cost accessories required).
22. You can locate the spot where the fish were biting.
23. You can amaze your friends.
24. You can find your altitude within a hundred feet. (probably a lot closer)
25. You can find out exactly what time it is.
26. Did I mention they are fun?
27. You can easily go back the same way you came.
28. If something flies out of the window you can mark the spot and return to the exact place.
29. You can amuse yourself on public transportation.
30. You can use it to learn something about navigation.
31. You can remember the location of that place you only go to once a year.
32. It can be a great navigation calculator. For example it will tell you how far apart two places are if you know their locations.
33. Many units can tell you when the sun will rise and set.

Marine users can find plenty of uses from the above list plus a few more.

1. If someone falls overboard you can return and find him or her.
2. In conjunction with a chart you can avoid shallow water.
3. You can avoid submerged objects.

4. You can know where you are without any landmarks.
5. They are just as much fun on a boat.
6. You can find the next buoy in the race or the trip home.
7. You can see if you're drifting.
8. You can hook it to your fish finder.
9. You can see true speed and distance over the ground.
10. You can win log races (if they let you use it).
11. You can use it to drive your self-steering system.

Other uses for a GPS include:

1. Documenting the location of that neat photograph.
2. Having something useful to talk about on your home video.
3. Doing some amateur surveying.
4. Finding the altitude of your house.
5. If you have a combined GPS/cell phone your friends can watch you move around on the internet if you wish.
6. A couple of combined GPS/frs walkie-talkies can be used to keep contact and help if one person gets lost.
7. Helps you justify the palm top you wanted to buy.
8. Instantly find your location on an electronic map.
9. You can use the backlight for a flashlight in a pinch.
10. Figuring out where the cruise ship or airplane is.
11. Setting your clock.

There are plenty of other users as well. These include aviators, hikers, bicyclists, motorcyclists, skiers, joggers, fishermen, hotair balloonists, kayakers, surveyors, astronomers, parachutists, and lots more.

...And, how about hobbies and Games

A GPS can be used to enhance your hobbies or to develop some new ones. Here is a list of some unusual items.

1. Some folks install a GPS in their model airplanes. Later, after the flight is over they can download a track and see where it went.

8

2. Ever wonder where your dog goes at night? Install a GPS on his back and when he returns you can find out.
3. You can search for confluences. These are spots on the earth where the lat/lon numbers all zero out to exact whole degrees. There are folks that think these are fun places to find! You can also look for government survey markers scattered around the world.
4. Other folks have set up hidden treasures (called geocaches) that can be located by GPS. Don't expect to find a million dollars though.
5. Maybe your next progressive dinner won't include any traditional addresses but only lat/lon coordinates. No cheating by saying whose house it is.
6. Some cameras can include location data right on the picture or perhaps you can just take a picture with your GPS in the foreground.
7. Golfers can use a GPS to display the distance to the pin. Useful for picking the right club to use.
8. A hunter might find that deer stand he/she used last year.

What about on your job?

Certainly many jobs spring to mind when thinking about a GPS. For example a surveyor, a pilot, or a ships captain, but how about a few more unusual uses.

1. How about for a rural newspaper route? Perhaps the papers would even get delivered when you were sick.
2. How about a real estate agent that wants to show a house in a strange neighborhood without getting you and the client lost?
3. How about any job that collects field data? Wouldn't the location where the data was found be useful?
4. How about a farmer? Useful to manage those microclimates on your property.
5. Besides, why shouldn't work be fun too?

Introduction

Chapter 2
Theory

This chapter covers the theory behind GPS operation. It may be a little bit harder to read than some of the other chapters, but once you dig in you should find it rewarding. Understanding how your GPS operates can provide you with the knowledge that will tell you when you can trust it and why it sometimes behaves strangely.

How Your GPS Works

The GPS system consists of three pieces. There are the satellites that transmit the position information, there are the ground stations that are used to control the satellites and update the information, and finally there is the receiver that you purchased. It is the receiver that collects data from the satellites and computes its location anywhere in the world based on information it gets from the satellites. There is a popular misconception that a GPS receiver somehow sends information to the satellites but this is not true, it only receives data. So, just how is it able to compute its position?

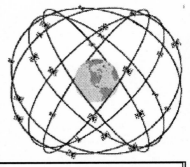

Figure 3 The Constellation

Geometric View

Your GPS receiver uses an elaboration of a technique that is tried and true and used by navigators and surveyors for centuries. Basically you use a known set of locations to compute your current location by taking fixes on the known sites. As a marine navigator you might take bearings (compass sightings) on existing locations and triangulated these on a chart to compute a fix on your location. Once you have a compass bearing you can draw a line through the known location and you know you are somewhere on that line. Do the same thing to a second point and the two lines will intersect. This is your position. If you try a third point it should intersect at the same place the other two lines intersect. Usually however, because of imprecise sightings, it intersects both lines at slightly different points thereby forming a small triangle. You are somewhere inside that triangle but you don't know exactly where. If the triangle is small enough you consider it good enough, otherwise you need to take another sighting. Accuracy is determined primarily on your ability to get and plot an accurate bearing as well as the geometry of the known sites available. This means that if the sites are very close together you will get poorer results than if they are at some angular distance apart. What you would really like were two sites that were 90 degrees apart for best accuracy.

The GPS receiver uses a slightly different approach. It measures its distance from the satellites and uses this information to compute a fix. How can it measure distance? Well it really measures the length of time the signal takes to arrive at your location and then based on knowing that the signal moves at the speed of light it can compute the distance based on the travel time. However, unlike the known sites of the olden days, these sites are moving. The solution to this problem is to have the satellite itself send enough information to calculate its current location relative to your receiver. Now, armed with the satellite location and the distance from the satellite we can expect that we are somewhere on a sphere that is described by the radius (distance) and centered at the satellite location.

By acquiring the same information from a second satellite we can compute a second sphere that cuts the first one at a plane. Now we know we are somewhere on the circle that is described by the intersection of the two spheres. If we acquire the same information from a third satellite we would notice that the new sphere would intersect the circle at only two points. If we know approximately where we are we can discard one of those points and we are left with our exact fix location in 3D space. Now, what would happen if we were to acquire the information from a fourth satellite? We should expect that it would show us to be at exactly the same point we just computed above. But what if it isn't? Before we can answer that question we need a little more background.

A more basic question is, "How does the GPS know the travel time so that it can compute the distance?" The satellite sends the current time along with the message so the GPS can subtract its knowledge of the current time from the satellite time in the message (which is the time that the signal started its descent) and use this to compute the difference. For this to work the time in your GPS must be pretty accurate - to a precision of well under a microsecond. The satellite itself has an atomic clock to keep the time very precisely, but your unit is probably not big enough nor expensive enough to have an atomic clock built in, so your clock is likely to be in error! For this reason our assumptions about the distance calculation are likely to have considerable error and the fourth satellite fix will reveal this to us. However, if we assume the error is caused by an error in our clock then we can adjust our clock a little and recompute all four solutions, continuing to do this iteratively until the error disappears! We will then have a good position fix and as a side effect we will also have the correct time to about 200 nanoseconds or so. One of the applications of GPS technology is to provide the correct time even when we don't care about our position.

Maintaining the fix means that we need to continuously recalculate the information based on the moving satellites. Once we have a number of fixes we can derive much more information than just location data. For example a GPS can compute the travel direction (compass heading) by comparing current location to previous location. Similarly the GPS can keep track of travel distance, compute speed, record travel time and other valuable data.

This view is simplified. In addition to the data already mentioned the unit uses Doppler data from the moving satellites, almanac data to figure out the approximate positions of all the satellites, and ephemeris data downloaded directly from the satellite that can be used to compute its position in the sky. For a more detailed look at this information you should read the section on obtaining a fix. Similar to the geometry problem we had in the older system of taking bearings on fixed sites, the satellite geometry has a significant effect in the accuracy of our final position. A unitless number representing this geometry is called Dilution Of Position, DOP and is used by the GPS in determining which of the satellites available represents the best ones to use. The smaller the number the better the geometry.

Mathematical View

Another way to understand the operation of a GPS system is to look at the math that goes into calculating a position. From Pythagoras we have:

$$Prs + T + Es = sqrt\{(X - Xs)^2 + (Y - Ys)^2 + (Z - Zs)^2\}$$

Where X, Y, Z are the positions we are trying to find and T is the time error at the receiver. The terms Xs, Ys, Zs are the satellite positions that can be calculated from ephemeris information sent from each satellite. The Es term is a lump sum of all the modeling errors considered by the GPS. These include such things as troposphere and ionosphere errors, clock errors from the satellite and any other error the GPS receivers thinks is significant enough to model. Prs is the approximate (pseudorange) distance from the receiver to the satellite. Since we can calculate the pseudorange and satellite positions independently and we can factor in modeling information from hardcoded data we are left with four unknowns, X, Y, Z, and T. Therefore we need 4 equations to solve for the 4 unknowns. Mathematically this is a standard least squares problem. One approach is to use guesses of our current position to calculate delta's from what we would expect and then iterate towards a converged solution. This is the reason that the unit requires an estimate of our current location to compute our position. Once we have the delta's down to an acceptable level we have a solution.

In actual practice a Garmin receiver calculates a set of equations with 7 unknowns. In addition to the 3 positions and time they have added the Doppler data dx, dy, and dz which represents the relative speed between the satellite and the receiver. These terms are needed because our solution is based on moving objects and dx and dy can be used as part of the receiver velocity calculation (dz is discarded). Four equations will compute a full 3D solution but new 12 channel Garmin units can use additional satellites to perform an overdetermined solution that will offer more accuracy. Older multiplex units pick the best 4 satellites based on their DOP. As satellites move out of view or get blocked from the receiver's view by buildings, trees, and other objects the receiver will switch to other satellites to maintain a location fix. If the number of tracked satellites drops to three then a

3D solution is no longer possible and the receiver will use the last available altitude and compute a 2D fix for horizontal position.

The Other Two Elements of the GPS System

In addition to the receiver we must have a set of satellites in the sky and a method of updating the data in each satellite. There are full time land based sites that monitor the various satellites that are often referred to as Space Vehicles, SV's. These land based sites check the health of the SV's, check how close they are to their optimum orbits, check the clock accuracy, and send adjustments as needed. The land based sites are located a precisely known positions so that they can verify the operation of the satellites.

The satellites are travelling around the world 11,000 nautical miles high in carefully controlled orbits at a speed that means they will make a complete orbit twice a day. Each orbit takes 11 hours and 58 minutes, so like the stars they will seem to drift 4 minutes a day. The complete constellation consists of a minimum of 21 SV's and 3 working spares. Currently there are 27 total satellites in the sky and it is possible that there could be as many as 31 or 32. There are 6 orbits with multiple satellites in each orbit as depicted in the drawing at the beginning of this chapter. Each orbit is inclined 55 degree from the equator and thus there are no orbits that go directly over the poles, but certainly a great many orbits can be seen from the poles or anywhere else on the earth. The goal of the system is to at all times provide at least 4 satellites somewhere in the visible sky. In practice there are usually many more than this, sometimes as many as 12.

Each satellite contains a supply of fuel and small servo engines so that it can be moved in orbit to correct for positioning errors. With update control from the ground units it can maintain an essentially circular orbit around the earth. It also contains a receiver to get update information, a transmitter to send information to the GPS receiver, an antenna array to magnify the weak transmitter signal, several atomic clocks to accurately know the time, control hardware, and photoelectric cells to power everything.

More Detail on Calculating a Receiver Position

The steps involved in calculating a position are:
1. Sync with an available satellite and download the navigation information. (See the section on obtaining a fix for more details.)
2. Convert the messages to internal format for calculation. These include clock information, ionosphere data, and ephemeris (orbit) data.
3. Calculate the exact satellite position. This will include both the elevation and azimuth data so we can apply tropospheric modeling corrections that are dependent on how far above the horizon the satellite is.
4. Calculate the pseudorange data and then correct for ionosphere and other modeling errors. (Note that consumer units may not fully compensate for ionosphere or tropospheric errors.)
5. Repeat these steps for each available satellite. Garmin will initially attempt to find 3 SV's starting directly overhead and compute a 2D fix using the previous fix altitude (or data input by the user).
6. Correct the SV position for earth's rotation based on the time it takes for the signal traversal using the pseudo range data. (If the internal clock is close this can be done once, otherwise it will have to be repeated after the receiver position is computed.)
7. Correct using differential data if available. (This may have to be done after the initial position is computed as part of the refinement step if the internal clock isn't accurate.) If the differential station is near the GPS receiver it will be able to skip the corrections for modeling errors since this is part of the correction data available. Using dgps corrections leads to accuracy considerably beyond the capability of a standard receiver.
8. Calculate the initial receiver position as described in the prior section.
9. Convert the data based on whatever datum and grid system chosen by the user and display the answer on the position

page. Altitude is also corrected for geoid height prior to display.

10. Add in the leap seconds and time offset from UTC time to the computed time data and then convert it for display.

11. Refine the position based on additional satellites and the correct time to obtain a 3D fix and subsequently improve the fix based on choosing SV's with a better DOP, applying an overdetermined solution, etc.

Getting A Fix With Your Garmin

The discussion that follows includes conjecture based on research and knowledge gained through the use of Garmin devices. It does not represent any inside information from Garmin and may not represent the proprietary methods used in the Garmin design. While this data is based on Garmin receivers, the methods described are similar to the methods used by any GPS receiver and this discussion should be useful to anyone trying to understand how a GPS works.

Before you can actually use your GPS receiver for navigation you must first obtain a fix on your current position. Generally this happens automatically each time you turn the unit on. Once you press the on switch the unit will run a set of internal diagnostics and then switch to the status screen (except for the emap), with an intermediate stop on mapping receivers to display a disclaimer page. Next it will retrieve information for the necessary satellites and once it has computed a position fix it will switch to the position screen. While it seems to be doing the same thing each time it starts there are distinct differences in obtaining a fix depending on the how long it's been and how far you have traveled since the last time you turned it on. Obtaining your first fix is sometimes referred to as initializing your receiver. There are several methods of initializing a receiver from seeding it with information (called EZinit by Garmin) to fully automatic self-initialization using auto-locate.

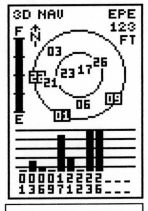

Figure 4 - 12XL
status screen

The status screen for the 12XL is shown above. It is similar to the status screen information on all Garmin units. One of the first things to catch your eye when the status screen appears is the two concentric circles with numbers seemingly located at random over the top of them. These represent the expected locations for satellites at your location and time. The outer circle represents the horizon and the inside circle shows a 45-degree angle. Everything inside the inner circle is overhead. Knowing the approximate location and satellite number for the expected satellites aids the unit in determining your first fix. This implies that the unit must already think it knows your approximate location and the approximate time. The approximate location of all of the satellites is stored in the machine in, what is called, the almanac. The almanac provides the data you see on the screen and aids the GPS receiver by letting it know which satellites are likely to be available. The almanac data is usually good for about 3 months and is updated automatically when the unit is in use. However if you leave the unit off for over three months it will have to re-collect much of this data before it can obtain a fix.

The Garmin emap will display this information but does not automatically switch to the status page. You will need to select "GPS info" from the main menu. The Garmin basic etrex displays very little status information on its satellite screen. Only 4 satellites are shown and they are not displayed with any geometric information. As the fix progresses on the etrex a light squiggled line will appear as satellites are found and this line will become darker as the ephemeris data is acquired. To understand how this really works you should change the basic etrex to the advanced SkyView page. The advanced sky view page shows the same information that is available on all of the other Garmin receivers.

(Note that some of the data in this chapter assumes a general understanding of how a GPS works. If you need this knowledge please read the section above on this subject.)

Cold Start

The first time you turn you unit on each day it must perform a cold start. To perform a successful cold start it must have a *current almanac*, a reasonable expectation of its *current location*, and a reasonable idea of the *current time*. Given this data the only thing it needs in order to calculate a fix is the precise location of 3 (for a 2D fix) or 4 satellites (for a 3D fix). The data it needs is called ephemeris data and it is transmitted every 30 seconds by each satellite in the constellation. It takes 18 seconds to download this information because it is only being sent at 50 bps. Since your Garmin was turned on at a random time with respect to the satellite transmission it can take up to 36 seconds to download this information assuming no interruption. It would take longer if it had to be received in order but your Garmin is capable of reassembling the information even if it is received out of order. Once the data has been acquired a position can be computed and a lock obtained. Generally this can be accomplished in an amazing 45 seconds from the time you first turned on the unit. The 45 seconds includes the time to gather ephemeris data and the time to compute the fix. After a lock the data will be updated once a second.

The older multiplexing receivers cannot achieve such a rapid initial lock. The reason is simple; they need to download ephemeris data from 3 satellites but cannot do so in parallel. So it takes 90 seconds to download the data from 3 satellites and thus you can expect the first 2D lock in about 2 minutes. After another 30 seconds or so you will usually be able to obtain a 3D lock as it adds in the additional data for the fourth satellite. If you have a unit that starts with hollow bars and darkens them later you can watch the progress on these units. Once a satellite ephemeris data has been collected the bar will turn dark. Since data can be collected in multiplex mode and out of order it is quite possible for several bars to turn dark at essentially the same time. Since the unit knows it needs 3 satellites to compute its first fix it will concentrate on getting the information from 3 high satellites simultaneously. After a 2D fix is produced it can keep track of those three satellites while attempting to add a fourth to produce a 3D lock. Ultimately the unit will try and track up to 8 satellites but will only use 4 in a solution. There is no indication about which 4 it will use at any particular moment but it changes them as needed to attempt to produce the most accurate fix.

The newer 12 channel parallel units can gather data simultaneously from up to 12 satellites. There is no time penalty in doing so. While the G-III family, emap, and most other Garmin units continue to indicate the collection of ephemeris data by turning the status bars solid the latest release of the G-12 family and G-II+ no longer uses the hollow bar vs. solid bar indication to indicate this. Instead solid bars are used to indicate exactly which satellites in the constellation are currently being used to calculate the fix. Therefore none will turn solid until the first fix is obtained. A feature of all of the 12 channel units is the ability to calculate an overdetermined solution where more than the required 4 satellites are actually used to compute the solution. Using more than 4 satellites can improve the error and can minimize the shifts that sometimes occur when a GPS needs to switch which satellites are being used to compute the solution.

Note that the hollow/filled bar feature is not present on all units. Most notably it is missing from the 12CX color unit and some of the really old multiplexing units.

Warm Starts

If you turn your unit off and back on again you will notice that it is able compute a fix much quicker than it did the first time. This is because it saves the ephemeris data from the last cold start. Saving this critical data can enable your unit to compute a fix in 15 seconds or less. It needs only to verify that the data is still valid and that the satellites that it has data for are still available in the sky. The ephemeris data from a satellite is updated every 2 hours at the top of the hour but it is considered to be valid for up to 4 hours. (If the circular orbit is projected to be really close to a circle they might declare 6 hours.) Considering that a satellite makes a full orbit in just under 12 hours and the earth is moving underneath then 4 hours represents most of the time it spends in transit overhead.

Garmin handheld receivers look at the time stamp and use this data when it is valid which permits warm starts whenever possible. Of course in several hours the satellites you were using may have drifted out of the overhead position and may no longer be available, but at least some of the time a warm start can be achieved. Note that the warm start does not require the same solution, it just requires that 3 or more satellites for which you have ephemeris data are still present in the sky and can be used to calculate a fix.

EZinit Starts

The basic etrex does not support EZinit starts. Please skip this section and go to the autolocate discussion if you own an etrex. The basic etrex will attempt a local fix and if this fails it will ask you some questions to determine if an autolocate is required. The EZinit feature is supported on etrex venture, legend, and vista.

If you have moved 500 miles or more since the last fix you will probably have to do an EZinit. In this case you will seed the unit with something close to your present position to aid in its ability to compute a fix.

This is done on non-mapping units by selecting the state or country you are currently in. This information can be reached by pressing the enter key (menu key) while the unit is trying to acquire satellites. You can then scroll through the list of countries and states. Once you have selected the starting location the unit proceeds as if it were a cold start. (If the altitude is greatly different you may want to help it along by manually entering an altitude that is fairly close as well.) On these Garmin units you may also initialize their position by changing the lat/lon position that is shown on the position page. While the unit is acquiring its position this is a changeable field on most units and may be selected just by using the arrow keys to select the field.

On Garmin units with built-in maps the map position can be used to seed the current location. Note that these techniques do not have to be exact since getting within 500 miles is close enough. To use the map, follow these steps: From the satellite screen, press menu and select Initialize Position, press enter. The map screen will appear and you can move the cursor to the desired starting point using zoom as needed to get you where you need to be. Pressing enter will set the starting point seed for EZinit.

Autolocate - Search the sky

The final method used by your Garmin enables it to find itself and get a lock completely without any outside help. In autolocate mode the unit does not use the internal almanac data but instead just uses brute force techniques to find the satellites. On any of the older multiplex receivers this can take 7.5 minutes to 15 minutes with a clear view of the sky for it to gather enough data to acquire a 2D fix. On the 12 channel parallel units the time has been reduced considerably since much of the data can be gathered simultaneously using the 12 channels. A 3D fix under these conditions can usually be obtained in about 3 to 5 minutes. Note that while a fix may be obtained, this is not sufficient time to reload the full almanac. The almanac requires 12.5 minutes to download from the satellites. For this reason, if you have to do an autolocate you should continue to leave the unit on while it regathers its full almanac, or you may have a similar problem the next time you turn it on.

If the unit is doing an autolocate and it fails to produce a 3D lock it may prompt you for the altitude reading. You should enter something as close as possible to what you might guess it to be. This usually happens when the unit has a restricted view of the sky such as out the side window of an airplane.

You will likely need an autolocate if you have let the almanac data get out of date by not using the unit for 3 months or more. You may need an autolocate if the internal clock is totally wrong although usually a EZinit will work in this case. Since the etrex, 76 family, GPSV, and emap do not have an internal battery backup this can occur on those units if the main batteries are left out for several weeks. Finally you may need an autolocate if you are somewhere on or over the ocean without a clue as to where you are. Under some conditions where the unit just refuses to lock you might want to use the undocumented clear unit command (See the chapter on undocumented commands) to completely clear the unit of all information and do an autolocate to get it running again.

Completing the Initialization

As soon as you achieve a 2D lock the unit will switch to the position screen on units that have a position screen unless you have switched screens yourself during the initialization process or are using an emap. You can be navigating at that point but you may not have a 3D lock yet. On some units, attempting to change the altitude reading on the position page can check this. If it can be selected you still have a 2D lock. For older multiplex units, and the default settings on the newer twelve channel non-mapping units, the altitude setting is in the lower right display section of the position screen. It can be reached by pressing the up arrow key once. If the altitude data is selected when you push the up arrow key then you still have a 2D lock. It will remain selected until a 3D lock is achieved and then the entry will no longer remain highlighted. This can be a handy way to check for 3D lock status while using the position page for navigation. If the altitude is clearly wrong you can correct it while in 2D mode to improve the accuracy of the fix. On the mapping units the 2D altitude setting is only available from the menu and will be grayed out if you have a 3D lock.

Once the lock is established the clock on the position screen will be updated to reflect the exact time. Internally the unit uses UTC time without leap seconds but it can be set to display local time with leap seconds added automatically. While the initialization would seem to be complete as soon as a fix is obtained the GPS will continue to find and collect data on other satellites in the background while you are using the unit. This activity continues so long as the GPS is turned on.

Losing a lock

If you enter a long tunnel or go under a dense tree cover you may temporarily lose your fix. To help overcome these problems Garmin has designed in a dead reckoning circuit. For temporary loss of signal under 30 seconds the unit will just continue to use its current heading and speed to project your position. If the signals are reacquired during this time the unit will adjust the position to correct for any errors in the projection and just continue on as if nothing had happened.

Loss of signal longer than 30 seconds triggers the "poor GPS coverage" or "weak signals" message. This message appears in the message window and continues to be displayed on the status screen. In addition anywhere that track and speed are displayed the fields will be replaced with '——-' to indicate that there is currently no fix available. Once a fix is re-obtained the track and speed data will return. Any break in the track caused by a loss of signal will also be recorded in the track log by showing a break in the visual log and a discontinuous log will be generated. Note that the projected log entry from the dead reckoning will be backed out of the log and the display.

The reacquisition time for a Garmin GPS is well under a second so if the satellites become visible again an almost immediate lock will ensue.

Loss of a signal for a considerable period will trigger a message to re-init the unit. If you know the cause is just a temporary loss of satellite coverage then just answer the re-init choice with "continue acquiring" to cause the unit to look for a longer period. If the unit determines that there are no satellites available after about 10 minutes it will shut itself off to conserve batteries.

Notes and Tips

- If you have changed altitude since the last time the unit was used you may get a quicker lock if you manually enter the altitude yourself during the time the unit is attempting to acquire satellites. Not only will this help in the acquisition but the 2D fix will also be more accurate. A large error in altitude will cause a large error in your fix location or even an inability to compute a fix.
- On the older multiplex units you may find that the unit seems unresponsive when you turn it on. If it fails to find a satellite, as indicated by at least one bar appearing within 10 or 15 seconds, then shut it off and try again. For reasons that are unknown to me this will often kick it into working.
- If the unit seems to have trouble finding a particular satellite it may be that there is an obstruction, such as your body, preventing it. Orient the unit North and study the status map. Look for obstructions blocking the satellites you need and turn your body or move to attempt to find a more suitable location.
- For best results, particularly on older multiplex units, be sure the antenna is oriented properly. The patch antenna should be horizontal while the bar antenna should be pointed straight up. This is much less important with the newer 12 channel units.
- There are times where an older multiplex receiver may grab all of the satellites and then seem to forget that it was supposed to compute the fix. Under these conditions you can try interrupting the signal with your hand to force a recalculate or you can turn the unit off and back on.

24

Turning the unit off and back on without an intervening fix causes these older multiplex units to erase the current ephemeris data and to collect it again.

- The specified lockup times mentioned in the text above assume the unit is stationary during the initialization. Moving during initialization seems to lengthen the time it takes to acquire a fix. The exact reason is unknown but likely has multiple causes. Certainly any momentary disruption of the signal such as tree cover or an overpass will cause the unit to have to re-gather the data it was currently obtaining. It may also be that it is related to the fact that the unit must do an iterative solution using the Kalman filter as part of its initialization step and changing one of the Kalman filter inputs lengthens the convergence time for the solution.

- A full almanac is transmitted every 12.5 minutes. This includes data that is used to supply the leap second data. If your unit seems to be off a few seconds all of the time it is likely it hasn't been able to update the leap second data from the satellites. Be sure that the unit is on for a full 12.5 minutes from time to time. If the leap second changes while the unit is running you will need to power it off and back on to see the results of the leap second change.

- It is possible for the status message at the top of the status page to indicate that the current fix is unusable. The message indicates that you should power the unit down and re-init. This will only happen rarely and you could use your unit for years and never see this message. It means the unit is confused about the fix typically caused by receiving conflicting information that won't result in a stable fix. One cause of this is intentional jamming that could be performed by the military as part of their testing efforts. The only recovery is to do what the message says.

Detailed Discussion

Locking onto a set of satellites can be a formidable task. First, all of the satellites are transmitting on the same frequency, 1575.42 MHz.

And second, the power output is very low (about 500 watts effective radiated power from 21000 km in the air) so that the received power is about -160 dbW, which is just barely above the power of the surrounding background noise and often below the background noise level. Imagine a hen going out into the chicken yard and hearing 100 chicks all crying at the same time. How does she know which one is hers? She finds her own chick because she recognizes its voice. A GPS receiver works in a similar fashion. It knows which satellite it is trying to receive and it matches a predefined code with one that is sent by that satellite. Once it makes a match it has found the correct satellite. At this point it will lock to that signal and display the strength bar on the satellite status page indicating it has synced to the satellite and how strong the connection is. Next it can begin downloading the ephemeris data.

Each satellite sends its signature every millisecond. A signature consists of 1023 bits of unique data that has been chosen specifically to aid in discerning this satellite from all of the others. There are 32 such codes defined. The receiver uses the almanac data to estimate the position of the satellite it is interested in and to predict its Doppler shift speed. It then uses its own location and time to attempt to line up a copy of the 1023 bit code to exactly match the code from the satellite. If it can't line it up exactly it will shift time or clock frequency slightly to get it to line up. Once it gets a line up with the satellite it was seeking it can search for the HOW (Hand over word) sequence to get in word sync to retrieve and decode the information. The data itself is modulated at a 50 Hz rate on top of the signature by using the signature as a carrier. There are 25 frames of data that is divided into 5 subframes of 300 bytes each. A frame is transmitted in 30 seconds, thus each subframe takes 6 seconds to transmit. The first 3 subframes in each frame contain the same data, which is why the satellite specific data can be obtained every 30 seconds. The First subfield contains health and accuracy data as well as corrections for the satellite clock. The next two subframes contain the ephemeris data. The final 2 frames contain all of the other data, such as almanac data, that is of less importance in obtaining the first fix.

Since the data the GPS receiver wants is contained in 3 of these 6-second frames the collecting of this data is the first essential task of the receiver. Once locked on it needs the HOW sync byte and it could take up to 18 seconds to get it if we suppose the HOW for the 3rd subframe was just ahead of our lock. In this case we will have to clock through the 3rd subframe, notice that we are on the 4th subframe and wait for the first subframe to roll around again. Under this condition it will take 36 seconds to get the data. Suppose the first HOW just went by; now we will find the 2nd subframe and will go ahead and gather this data as well as the 3rd subframe. We will then clock through subframes 4 and 5 and get the 1st one, another 36-second case. A best case scenario is one where the first subframe HOW appears immediately after the sync and will result in the data being present after 18 seconds.

Once sufficient satellite data has been collected it will be combined with data from other satellites to solve a set of simultaneous equations involving 7 unknowns. These unknowns are X, Y, Z, dx, dy, dz, and t. The speed vector dz is discarded leaving a 3D fix, horizontal velocity vectors, and the correct time. If only three satellites are available the Z (altitude) setting will be used from the unit itself based on the previous fix or data supplied by the user. The nature of the method used in the solution also uses the current X, Y, and Z positions as an initial value for the calculation. Garmin doesn't release the exact techniques and algorithms that they use to perform this computation but it seems clear it is similar to the Kalman filter methods where current information is used to seed the new information in an iterative form. Iterative techniques are often used in computer programs to solve complex problems lend themselves well to continued use to update the data while the GPS is in use.

On 12 channel parallel units Garmin has developed a technique that permits them to calculate an overdetermined solution by using more than the standard 4 satellites. This can lead to more accuracy, reported to be on the order of 15%, and permits a smoother transition when a satellite drops out of view for any reason. The G-12 family indicates exactly which satellites are currently being used in the calculation by showing black bars. Satellites may be dropped from the solution for a number of reasons, such as the detection of significant multipath reception on one satellite. Other 12 channel Garmin receivers use black bars to indicate only that ephemeris data has been collected and do not show the exact satellites being used at a point of time although this data is present in the NMEA sentences on the computer interface.

Theory

Chapter 3
The User Interface

This chapter covers the main differences in using the different Garmin units. It focuses on the User interface while presenting an interface philosophy that should permit a user to be able to translate user interface differences between models to the idea behind the interface. The goal is to permit a user to adapt to various models and know how to do a particular function even though a model may be using a different philosophy from the one the user originally learned.

Working with the Keypad

This section covers the philosophy of the Garmin interface and the specifics of most of the handheld units. The emap and etrex are sufficiently different to have their own sections but the overall philosophy is the same. A quick look at the keypad is usually enough to tell the user which of the various interfaces philosophies is used on a particular unit. The emap uses a keypad that is similar to the other handheld units while the etrex uses keys that are all on the sides of the unit. The specifics of the emap and etrex user interfaces are covered in separate sections but the general discussion below applies to them as well.

As with many handheld electronic devices you will be working with your Garmin GPS using the keypad. In this case you have a very small keypad not much larger than a watch. Garmin handhelds have ten or twelve buttons that provide the full interface. Actually on later machines the 4 buttons that were the arrow keys have been replaced with a single rocker key that still provide the same 4 directions, while some of the latest units from Garmin use a miniature joystick that they call a click stick. The keypad on most older units looks like the diagram below: in out zoom keys not always present

```
     in out        zoom keys not always present
  goto  ^  page
  pwr <  > mark    called menu on the III and 76
  quit  v  enter
```

Other units have similar functions although the keys may be arranged differently. In particular some units have a "click-stick" which looks like a small joystick and performs the same functions as the arrow keys or rocker pad. In addition the stick may be "clicked" by pressing straight down which performs the same function as the **enter** key (covered in the next section).

Page Key

The display consists of a series of screens called pages and is very much like a digital watch display. One of the fundamental user interface features is the **page** key that rotates through the various display pages in the forward direction. (The **quit** key can be used to rotate through the pages backwards.) The display pages show GPS status, positional information about your location, map data, navigation information, routing information, and menu commands to access features that are not available directly from the keypad. As you might expect with so few keys the ones you have are used for multiple things. For example if the GPS alerts you to read a message then the page key doubles up as the key you press to read the message. You would then use the quit key to return to the screen you were using before the message appeared.

Quit Key

The **quit** key is how you back out of what you were doing. Thus it will move you the opposite direction from page or it will cancel a command. Successive presses may be required to get back to the starting point. On the emap the key is called esc.

The Object-Oriented Interface

One of the more interesting features of the Garmin interface is that it emulates the object-oriented interface often found with software products that use a mouse. Basically graphic oriented products often let you move the mouse to some object on the screen and just click on it to perform a command or function.

Enter Key

Your Garmin emulates this behavior using a select and execute command sequences that combines the use of the arrow keys with the **Enter** key (or sometimes the menu key on units so equipped) to perform the "click" operation. If you use the arrow keys on any of the pages on you Garmin GPS you are likely to find that something on the screen automatically becomes highlighted. As you continue to press the arrow keys you will notice that the object that gets highlighted changes very much like when you move the mouse over objects on the screen. For the Garmin units almost anything on the screen can be an object. For example a title of data can be an object or the data itself can an object if the user can change it.

If there is only one thing that can be done with the object then selecting it will automatically bring up the screen that lets you work on the object. It there is more than one thing that can be done then the selection will bring up a choice list or menu of commands that you can use. If you have a choice then use the arrow keys to highlight the choice you want and press the enter key to select it. Or you can use the quit key to cancel the command. While an object oriented interface is unusual in a small handheld device the power it brings is appreciated by most people who use it. It provides for less key presses than would be needed by other interface techniques which caters to the power user. The key to understanding this interface is just to use the arrow keys and see what you can select. Depending on what else is going on you may be able to select different things at different times. For example when using the simulation mode available with your unit you may find that you can select additional objects or some objects may be selectable when you first turn the unit on but can no longer be selected after the unit figures things out on its own.

Using the arrow keys is circular. The down arrow key attempts to move down but if you are already at the bottom then it circles around and you end up at the top. Similarly the up arrow key will also circle. This fact can be very useful if you start at the top and notice the object you want is near the bottom of the screen. Moving up will often get you there quicker than the more traditional way. The left and right keys may do similar things to the up/down keys or they may attempt to highlight items that are logically left or right on the screen. They are also circular but will tend to move up or down as well when the edge of the screen if encountered. Left moves down while right moves up.

The menu interface

While the object interface is powerful there are certain things that are better accomplished using a menu interface. There are three fundamentally different approaches to the menu interface that are used on Garmin receivers. This becomes obvious when you notice that some receivers have a dedicated menu button on the face of the unit while others seemingly don't have any menus at all. A third group of units has no menu button but every screen has a banner at the top with menus always available from the banner. These units are similar to the ones with a menu button in that the interface is very menu oriented.

Menu Key

All of the units in the Garmin III family which includes the G-III, the G-III Pilot, the G-III+, and the 12Map, as well as the 76 family and emap have a menu key where the mark key would normally appear on the other Garmin units. There is a significant difference in the way that you work with these units because of the menu-oriented interface but all of the functions are the same. Often the screens and commands are simpler and more straightforward which appeals to some users. However, there are cases where the fundamental object oriented interface is also present. There are two basic menus available to the Garmin units whether explicitly or implicitly defined. There is a local menu associated with each screen that contains commands intended for that screen and there is a global menu that contains all of the machine configuration items and commands that are not specific to one screen. On units with a menu key pressing the menu key once brings up the local menu. Pressing it again without any other keystrokes brings up the global menu. There may also be submenus available either directly from the menu or from a screen that appears as a result of selecting an item from the menu. In some cases the submenu is reached by pressing the menu key but in many cases this is accomplished by providing a tab at the top of the display. These look like file folder tabs and behave similarly. Highlight and select the tab folder to bring it and its contents to the front. Note that there may be more tab folders than can be displayed so you may have to use the left/right arrow keys to scroll sideways to see them. In all cases once the menu appears the arrow keys are used to highlight an item and the enter key is used to select that item or the quit key (esc on the emap) cancels the operation.

The etrex legend and its siblings have a banner on every screen with the screen name on it and two small boxes on the right side. One is a menu that permits selecting from the available pages (including the main menu) and the second is the local menu. Use the "click stick" to highlight the menu (left) button and then click to bring up the local menu. The right button can be used to select a screen providing direct access as an alternative to paging through all of the screens one at a time. A double click on this button will permit toggling between two pages of your choice. Note that the click stick performs the selection movement and may be "clicked" to provide the **enter** function.

All of the other Garmin handhelds also have a menu interface but it is a bit subtler. If you press the enter key when there is no object highlighted on the page then you will see a local menu. (Or if there is only one thing that can happen you will execute that command.) Once the menu appears you can highlight the desired command with the arrow keys and select it using the Enter key, or cancel with the quit key, just like you would using the G-III menu system. In addition there is a global menu. Since there is no menu key the global menu appears as just one of the pages in the standard page rotation. So if you hit the page key a few times you will reach the global menu. Submenus are a bit more problematic on units without a menu key. What Garmin has done is to divide the screen into two portions whenever there is a need for a sub-menu. The main screen data plus a menu selection at the bottom of the screen. Most of the design differences in the screen displays between the G-III family and the other units is because of this one difference.

In addition to the local and global menus there is also a banner menu on the map page of some units. Banner menus appear across the top of the screen and contain choices that are specific to that screen. They are easy to recognize because they participate in the highlight mechanism used to select other objects on the screen. Moving the highlighted object using you arrow keys will rotate between screen objects and the banner menu. There will never be a case where nothing is highlighted, which is how you can recognize the banner menu. Selection is done as always using the Enter key. The newest etrex units with the click stick have a banner menu available on all screens. It is used as an alternate way to jump from screen to screen and to access the local page menu. The banner may also display unselectable information that is useful to the user.

The Function Key Interface

So far the discussion has included the use of the arrow keys, the page key, the enter key, the quit key, and the menu key (if present). It would be possible to design a GPS with only this set of keys, or even less, however Garmin has singled out some important functions and provided special keys to make them easier and quicker to use. These keys, like the ones we have already discussed, often have multiple uses as well. In some cases the secondary use is reached by pressing the same key a second time in succession like was done with the menu key as described above. In other cases holding the key down for a longer period of time accomplishes the second function.

Power Key

One of these keys is the **power on/off** key. This is the red key shown above on the left center of the control panel. It contains a small icon showing a light bulb in its center that points out its secondary use. If you press the power button and the unit is already on the switch will turn on the backlight. On some units pressing it multiple times will toggle through a selection of available lamp brightness' while other units only have a single lamp function. To conserve battery power the unit has the ability to automatically turn the lamp off after a preset period. If the lamp has turned off by itself then any key depression will light the lamp and you will need to press the key a second time to perform the intended function of the key. To turn the unit off you must hold the button down for several seconds. The 12CX uses this button to bring up the contrast controls as well as for backlighting. The first time it is tapped it brings up the contrast form and the second and third taps control the backlight brightness. The GPS V also uses this button to bring up the contrast controls and lamp controls. The etrex models and emap have placed the power button on the side of the unit.

Goto Key

The **goto** key provides a quick way to set up the navigation information needed to use your GPS as a navigation device. You cannot navigate if you have nowhere to go. The goto key permits defining a destination. Once a destination is selected then the navigation features of your unit can help you to get there. As shipped, the Garmin units include one possible destination, their plant site. Perhaps they have concluded that everyone would want to visit them so they included this location in the default lists of destinations. Actually, it is more likely this is used as a test value for testing the units in manufacturing. Thus some units now also list their manufacturing site as a choice. As you begin to use your GPS you will find plenty of other destinations as well. Once you press the goto key, a list of possible destinations will appear and the unit will take a best guess as to the one that you might want. Here is another example of the object-oriented approach. If you have a destination highlighted on a list of waypoints or on the map screen then the GOTO command will automatically display this as the default choice. At the bottom of the screen is a submenu of other possible commands. All units have a Cancel GOTO choice and some of the newer units permit you to initiate a TRACKBACK from this menu. (More on trackbacks is in the chapter on routes.) The latest units permit you to jump to the submenu using the right arrow key, otherwise you will have to continually press the up/down keys until you reach the command. (Remember on the G-III these submenu items are on the folder tabs.)

In addition to the primary goto function there is a special goto function that is reached by pressing the GOTO key twice in a row. This automatically create a special waypoint called MOB at your current location. (Not available on the G-III Pilot or units without a goto key but is available on the 76 family using the nav key.) Pressing enter at this point will cause the navigation screens to begin to show navigation information to the MOB waypoint. MOB stands for Man OverBoard and is intended to help a boat return quickly to the spot where a person fell out of the boat. Most people find other uses for this as well.

The G-III pilot uses the secondary function of the goto key to perform a find nearest command by holding the goto key down for a second or more. This function paved the way for the dedicated find key that appears on the emap, newer etrex units and the GPS V.

The newer units do not support a goto function key but use an object oriented approach of selecting the waypoint and then selecting goto. On the 76 and Map76 this has resulted in the goto key being replaced by a nav key that is used to toggle navigation on/off, but the secondary mob function on this key is still present if you hold the key down.

Mark Key

The **mark** key is used to mark (save) your current location. On units with no mark key the **enter** key can be held down for a second or so and the mark screen will appear. Once the mark screen appears you can save the current location as a new waypoint. Waypoints are simply saved locations. They can be used as destinations or intermediate destinations on a route. They will be shown on the map screen when you get close to one of them. Waypoints are quite useful in helping you orient yourself to your present location or when navigating.

Page Key secondary functions

On the G-II family, the G-III family, and the GPS V the page key also has a secondary function. These units are capable of displaying their screens either vertically in portrait mode or horizontally in landscape mode. Holding down the PAGE key for a short period will cause the screen to switch modes. (You can also do this from the global menu but this key is a lot easier.)

The etrex has limited keys and uses the page key as a cancel key as well. If you are inside the menu system the page key is defined as cancel while its page function only really works if nothing else is selected. On any unit the page key may be the best way to bail out of a bunch of menus if you have moved deep within the menuing system.

Zoom keys

Some units also have two extra keys that are dedicated to **zooming**. These are used on the map page to change the scale of the map. They may also be used on the road navigation page to change the scale of the displayed road that, in effect, changes the scale of the cross track error. This function is implemented on those units having a cross track error display as part of the compass the compass page. On the etrex the zoom keys are on the side of the unit.

The find key characterizes most of the new third generation units. It indicates addition functionality is present that expands the waypoint idea to include other objects that can be 'found' in the database.

Find Key

The **find key** was new with the introduction of the emap handheld unit and copies the Street Pilot functionality. (The Street Pilot is Garmin's name for their larger automotive units.) It becomes important when you need to look up, or find, information on the map and demonstrates the focus of that product on mapping and locating information. In addition Pressing and holding the find key while navigation is active will jump to the goto waypoint screen or route screen as appropriate showing the active route. This provides a quick way to see the current destination since this is not displayed on the main screen and it provides more navigation data. All of the newest units include a find key.

The find key brings up a menu that permits finding any of the objects in the database of the unit. Which objects are available are determined by what maps may be loaded in the unit. Possible objects include waypoints (always present), cities, freeway exits, POI's (points of interest), street addresses, and intersections. Many objects can be listed either by their distance from your location or in alphabetical order. Some units will even let you set up a favorites list. The GPS V offers a list of recently selected places.

Generally the function keys can be depressed at anytime and will perform the function on the key. For example you might be doing something in the menus and suddenly find the need to mark a location. The operating system supplied as part of your unit is designed to permit this kind of interaction.

Basic Etrex interface

The basic etrex only has 5 keys and these are located on the sides of the unit where they are expected to be operated while the GPS is being carried in the palm of your hand. These are the power on/off key, the page key, an up key, a down key, and the enter key. These are shown in the diagram below:

```
     Up    <        > Page/Cancel
     Down  <

Enter/mark <         > Power/lamp
```

The page key is present to move from screen to screen just as in the other handheld units. It also doubles as the cancel key whenever you are within some submenu function. The power on/off key doubles as the lamp key just as it does on the other units. The enter key also retains much of its same meaning and even doubles as the mark function by holding it down for a long period just as it does on the Garmin III family and emap. It can be used to bring up a local menu on the map or pointer screen.

A unique feature of this unit is the lack of a keypad or arrow keys. Instead there is the much simpler to manipulate up/down key set. These keys have dedicated functions within the scope of the screen that is currently being displayed. These functions are:

- Status screen - perform contrast adjustment.
- Map screen - perform dedicated zoom function.
- Pointer screen - perform the function of rotating the display of one field of information. This is a kind of roller wheel display.
- Menu Screen - highlight the desired menu command.

There are very few dedicated keys on this unit. The functions performed by dedicated keys on other units are reached by accessing the main menu on this unit. The goto function is provided using an object oriented approach of selecting the object (waypoint) first.

Etrex: Legend, Vista, and Venture

These are new models from Garmin that share a case design with the standard etrex but are characterized with a new keypad area called a "Click Stick". This is a small joystick located above the screen display. It performs the standard functions of the rocker key and can be depressed straight down as well to do the enter function. With these functions handled by the "click stick" the 5 keys on the sides can be defined for other things. This leads to the following key arrangement:

Figure 5 Etrex Click Stick

The zoom functions are used to control the contrast on the status page and the find key is used to provide a new direct function for searching the database. This arrangement on the sides of the unit lends itself to handheld use but is a little cumbersome for use mounted in a car.

The Emap

The emap keypad arrangement looks similar to the keypad on other Garmin units except that it is below the screen area and, upon close inspection, there seems to be some keys missing. Indeed there are, the power and backlight keys are on the side of the unit instead of the top and they are separate keys. The backlight keys doubles as the key used to bring up the contrast menu by holding it down for a second or so. The keypad arrangement looks like this figure:

```
  in  ^  out    dedicated zoom keys
 find <  > menu
  esc  v  enter/mark
```

The key names on the keypad provide a clue that this unit is different from the other Garmin products. It focuses on using a map display and is designed from the ground up to be a mapping receiver that uses the functions of a GPS to provide information for the map display. When the unit is turned on you must acknowledge the warning screen (or wait for the time out) as with other mapping receivers and you are then shown the map screen. There is no page key or any method to rotate through other screens. Instead the navigation data that you may need is available from the main global menu (reached by pressing the menu key twice). The menu and enter/mark keys are in their traditional locations while the zoom keys replace the goto and page keys. The goto function key is not used on the emap being replaced with an object oriented approach. The esc key performs the function of backing out of any menu item (escaping) and if pressed and held down for a second it will totally back out from all menus, screens, and functions and return to the map screen. The find usage was covered above in the general section.

Entering data

To enter data into the GPS you must first highlight the field containing the data you wish to change. The full field will be highlighted; you then push the enter key. The highlighting will change to just a character to indicate selection. Now you will use the arrow keys to actually change the data. There are two kinds of fields, one

3. An object oriented interface where you use the arrow keys to highlight an object and then operate on that object in some fashion.
4. The menu system where you bring up a menu and select items or commands from the menu. There are usually two kinds of menus, the main menu that effects all operations and a 'local menu, which is unique for the screen being display. The main menu may be available by pressing the menu key twice or on units without a menu key it will be in the screen rotation. The local menu is always available on a key, enter or menu except when there is a banner.
5. Some or all screens feature a banner. If present it will be selected by the keypad or click stick like other objects. There are usually multiple objects that can be selected from the banner.

Data entry is accomplished using the keypad and selecting the characters by scrolling or menu selection. The enter key is used to confirm the data entry or the individual selected character on the etrex.

In the rest of this manual it will be assumed that you know how to use the interface and the various differences between the models. For example if the subject is about marking a location you should know that you can use the mark key on some units but may have to hold down the enter key on some others. Equally you should know how to access the global or local menu on your particular unit. If there are unique techniques that are model specific then they will be covered as needed.

The Display and System Setup

For the most part the display interface is fixed by the design of the unit. However on some of the newer units there are plenty of things you can customize to meet you needs. Some units are designed for use over the whole world and even permit customizing the language that the messages are displayed in. All Garmin units will work worldwide but the ones with built-in databases may be optimized for one area or another. For example the city databases may show more and smaller cities in one area and the maps may favor a given part of the world.

System Setup

The screen display customization for each screen will be covered in the chapter on display screens and navigation customization will be covered in the sections on navigation and coordinate systems but there is also customization that is just for the system itself. This customization is reached from the main menu under the title SYSTEM or SETUP and then SYSTEM. Once there you will see some information about the date and time. The date is set automatically and the time is also set automatically by the GPS system. You can change the display of the time by changing the entry marked OFFSET. This is a numeric field where negative numbers represent time west of UTC and positive numbers are east. PST in California should be set to -8:00 hours and during Day Light Savings time it should be set to -7:00 hours. You can also decide how time is to be displayed: in 12-hour AM/PM mode or 24 hour military time. On the etrex and emap you can also set automatic daylight savings time support and set the time zone using more traditional pneumonics in some cases.

Screen Contrast can also be set here or in the status screen (except for the 12CX, etrex, and emap) and there is a setting for the backlight. Turning on the backlight causes the unit to consume more current and thus the battery life suffers. For this reason the lamp will automatically turn off after a fixed delay. You can customize the length of this delay on this menu. Setting zero delay causes the lamp turn-off delay to be disabled. Most units also disable the delay automatically when you are using external power to operate the unit.

You can also set the mode of operation from this screen. These include normal, battery-save and simulation mode.

- Normal is the highest performance mode and is used for navigation.
- The simulation setting is only used when you do not actually need to track satellites and offers battery savings since the receiver portion is shut down. On some units this is called in-doors, GPS off, or demo mode. It will be

turned-off each time you power off the unit to avoid accidentally confusing the user on the next startup. Simulation mode has many extra features that are covered in the simulation chapter. You may be able to set this mode from the status screen also.

- Some machines and software versions also include a battery-save mode, also known as power save. Battery-save mode is remembered when you power off the unit. It works by noticing whether the unit is doing something unusual. If it was able to predict what will appear on the next sample from the satellites then it could have saved on batteries by not performing this sample. In battery-save mode the GPS decides how often to sample data based on its ability to predict the next sample. If you are standing still or moving in a straight line at a set speed then there is no need to sample data from the satellite as often since the GPS can predict what the answer is. If the GPS detects that something unusual is happening then it samples every second as in normal mode. Of course there can be a slight lag before the unit figures out that conditions have changed but this method works surprisingly well and under some conditions it can offer 50% greater battery life. Something changing means a change in direction or a significant change in speed and is similar in concept to the automatic tracklog detection.

On newer 12 channel units the battery save mode, also known as power save, will also limit the number of satellites being tracked. This is done by changing the mask angle to prevent the unit from attempting to track satellites that are close to the horizon. (Older multiplex units do this in all modes.) Solutions depending on these low-lying satellites are generally less accurate due to tropospheric delays. With a clear sky view battery-save mode can offer a significantly longer battery life with little degradation in service. In forest regions the accuracy suffers much more, particularly when traveling in vehicles. Waiting 5 seconds (3 seconds in older units) between fixes can cause a loss of lock in some cases and contribute to even longer excursions offtrack than a momentary loss in normal mode. If you are experiencing long straight tracks with sharp bends when traveling on a winding road the battery-save mode should be discontinued to maintain best accuracy.

Another setting that is user specified on some units is the Language setting under the SETUP menu selection. Here the user can specify which language they prefer to use. The G-12 family and the new etrex family support this feature and permits choices that include: English, Danish, French, German, Italian, Spanish, Swedish, Portuguese, and Norwegian.

The G-II family, the G-III family, and the GPS V, have a portrait and landscape mode option that can switch the display to optimize for use while being handheld or when placed on the dashboard of you car. This can be switched by holding the page key down for a few seconds or setting the preferred direction in the System Setup menu. Note that, in particular, the G-III family has independent customization settings for the portrait screen and the landscape screen. You can move information around for the two screen displays or you can even have different information on the two displays.

The GPS V has a battery selection option and WAAS selection on this screen.

All Garmin units have alarms that alert you to any unusual events. The alarms can be visual or audible. While all units can provide visual alarms only certain units can sound an audible alarm. The visual alarm is sounded two ways; a box appears alerting you of the presence of a message, and at night when the lamp is in use it will automatically light for alarms after its normal time out. Generally you can acknowledge an alarm by using the PAGE key to read the message or if you ignore it the alarm will reset if the condition that caused it goes away. On units with an audible alarm you can customize the tone usage on the setup page. You can set it to beep on key presses, only messages, or off.

On most older units part of the alarms are configurable from the ALARM entry on the setup page. Alarms associated with arrival at a waypoint can be turned off, on with a specified distance, or automatic. Automatic alarms occur when you are one minute away from getting to the waypoint at your present speed and direction. The etrex units have set the alarm to 15 seconds while the emap and GPS V automatically increases the alarm time based on the speed as shown in the table shown below.

Table 1

Speed (mph)	Alarm Time
0-30	15 seconds
30-45	30 seconds
45-60	45 seconds
60-75	60 seconds
70-90	75 seconds
> 90	90 seconds

There may also be a configurable CDI, or off course alarm that can be turned on or off. (CDI is explained in the navigation chapter.) Some units designed for marine use also have an anchor alarm that can be configured on or off as well as how far you are going to permit the boat to drift. There can also be proximity alarms, which are triggered by being within an individually configurable distance of a waypoint. The Garmin III family also has a visual alarm clock that can be turned on or off.

Units supporting WAAS will have an enable/disable setting on this menu. WAAS is covered in the dgps chapter. Selecting enable turns WAAS on. You should be in normal mode for this to work properly as WAAS needs to receive updates more often than permitted in battery save mode.

On many units all of system setup is on a single menu pick called setup. You then tab across to get the setup settings you need on mapping units or pick from a submenu on other units. Most of the system settings are described above but other settings are described in other chapters as needed. The time setting includes a display of the current time and is the only place that you can see the exact seconds for the time on the emap.

Batteries

Any standard AA batteries can be used in Garmin receivers. Some units have the ability to set the battery type in SETUP so that the battery gauge properly reflects the state of charge for the battery. For example in the G-12 family of units a fully charge set of NiCad batteries will only show 3/4 charge on the meter while the G-III family setup entry to support NiCad batteries would show full charge.

This setting generally only effects the meter and the low battery alarm, which is set to approximately 4.0 volts (1.9 volts on the etrex, 76 series, and emap), will work fine for either battery type although you won't get much warning for rechargeables. On some units the low battery warning is modified as well to give longer time. On one tested emap the momentary current drain of the audible alarm will shut the unit down at the low battery warning so there is no warning at all unless the alarm sound is turned off. There is more information on batteries in the "Getting Started" chapter.

All of the GPS units with the exception of the etrex, emap, GPS V, and 76 family also contain a backup battery. (These 3-Volt units and the GPS V do not need a backup battery since the data in these units is stored in non-volatile memory.) The internal clock on these units is run from a storage capacitor and will only last a week or so without batteries but this is not critical. A backup battery preserves critical user data when the normal batteries are removed from the unit. The backup battery has a life of about 3 months but it is rechargeable from the main batteries. If this battery runs down you will lose all user data and the almanac, which will greatly increase the time to obtain a fix. You will need to recharge it by leaving a set of regular batteries in the unit for about 24 hours with the unit off or on external power for a couple of days. It is best to keep a set of alkaline batteries installed at all times, even when you are not using the unit, to ensure that the backup battery remains charged. On the etrex and emap it is probably a good idea to remove the batteries if you are not planning to use them for a long period. Expect a longer lockup time the first time you reuse these units after storage. The almanac data will probably be out of date and the internal clock will likely be wrong, particularly if the batteries were removed from an etrex, 76, V, or emap. These units will keep the internal clock alive for a week or two without batteries. After that period, if a fix doesn't happen quickly, you will need to tell the unit that it was stored without batteries. This is a menu item on the status page that appears if a lock takes a long time.

Chapter 4
Working With Garmin Screens

Garmin units display a succession of screens that can be used to gain information about your GPS position and many other facts related to GPS operation and navigation. Generally you reach the various major screens in a rotation by hitting the page key to progress forward though the screens or using the quit key to move backwards through them. Other screens can be reached using the other function keys on the unit or via the menu system. The various models have slight, or major, differences in what is actually displayed on each screen and even different releases may vary the display somewhat. The figure 7 below provides a sample of screen contents taken from the Garmin GPS12XL version 4.00. It will be used as a guide for describing the various Screens. Note that while this chapter describes the screens themselves other chapters go into much more detail on how to use the screens to do various tasks. Be aware that while this manual refers to screens, Garmin calls them pages.

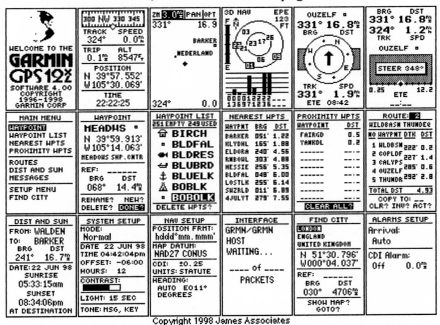

Copyright 1998 James Associates

Figure 7 12 XL screens

The Opening Screens

The first screen that you see when powering up your GPS is displayed while the unit completes its power on diagnostics. On most units it will display the version number for the software and hardware of your unit. (On some units this information can only be obtained using undocumented commands, holding down the enter key while pressing the power on button. Please see the chapter on using secret commands for more details.) This screen is depicted as the top screen on left in figure 7 above.

Garmin has a unique view of versioning as depicted on this screen for most units. (A few units have this information on another screen that is reached from the main menu or only available from special turn on sequences documented in the chapter on undocumented features) The first released version is called 2.00 as all Version 1 numbers are used for prototypes. The number after the decimal point is incremented when changes are made to the firmware in the unit. The integer portion of the version number is changed when major changes to the hardware are made. This means you cannot generally upgrade the firmware to a release where the integer portion of the version number is different from what you have. When the hardware is changed the firmware is usually changed to support the new hardware, which causes a major jump. The firmware version will be reset to .00 to signify this "new product". Occasionally the firmware version jumps by changing the first decimal digit thereby jumping over several incremental releases. This may signify a large change to the firmware and may include a change to the hardware as well. You may or may not be able to upgrade the firmware to this level. On early versions of Garmin handheld units the firmware was stored in read only memory and could not be upgraded without returning the unit to Garmin, however all recent units are now field upgradeable using software provided by Garmin for this purpose. In order to perform this upgrade you will need access to a Windows based machine and an interface cable available from a variety of sources. Details for this are in the interface chapter.

Once the power on diagnostics have completed you will leave the opening (welcome) screen. If you have a G-III family or other mapping receiver you will then go to the disclaimer screen which is lawyer talk for don't trust the maps. At the bottom of this screen may be the version number for the base maps themselves. They are stored in a ROM and are not updateable. You can ok this screen or wait until it times out. The next screen you will see is the status screen (G-III screen illustrated) or directly to the map screen on the emap. The newest etrex mapping units present yet a third screen warning you to be careful.

The Status Screen

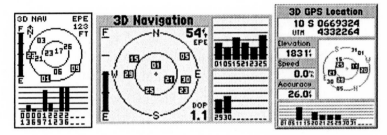

Figure 8 Status Screen

The status screen is the first screen in the main rotation. You can return to this screen at any time using the page or quit keys to bring it up in the rotation sequence. There are several pieces of information on this page about the status and quality of the fix that your GPS currently has. This G-12 screen is shown at the left. Most units will look similar to the G-12 figure however G-III family and G-II family units in landscape mode will show the status circle beside the status bars instead of above as shown in the middle of figure 8.

The emap also has a status screen as shown on the right of figure 8. It is not in the rotation since this unit doesn't use a rotation of screens. The status/position screen on the emap can be reached by going to the main menu (press the menu button twice) and selecting GPS info. The screen displays most of the information shown below and also displays information that is shown on the position page on other units. The emap has the largest display of any of the handheld units, which permits it to combine screens. The 76 series has a screen as big as the emap and permits it to offer status data and a choice of user customizable information on this screen.

The basic etrex and etrex summit have a status screen but the data on the main status screen only displays symbolic data that will indicate when you get a fix but little else. If you press the enter key a menu will appear giving access to an advanced screen. Selecting the advanced screen will display information similar to other Garmin units as described below.

Screen Contents

The upper left corner contains a message about the status of your fix. You may see messages like:

- Searching - Unit is searching the sky for satellites.
- AutoLocate - Unit is collecting new almanac data
- Acquiring - Unit is collecting data prior to its first fix
- 2D Nav - Unit has a 2D fix and is looking for a 3D fix
- 3D Nav - Unit is tracking normally
- 2D diff - Unit has a differential 2D fix and is looking for a 3D fix
- 3D diff - Unit is tracking fully using a beacon receiver.
- Not Usable - Unit was not able to compute a solution - Power down and try again.
- Poor Coverage - Unit is currently not able to compute a position. It will try and recover from this.
- Enter Altitude - You need to go to the position page (or use the local menu) and enter your guess of the correct altitude to help the unit initialize.
- Simulator - Unit is currently being used in Simulator mode.

Some of these status messages may also appear as popup messages while you are using the unit. In particular poor coverage is displayed when coverage is lost.

The upper right corner (left side on emap) has an entry for EPE (Horizontal Estimated Position Error). This is a report in Feet or Meters for the probable error as calculated by the unit itself. This is an indication of accuracy and is actually called accuracy on the emap. The number is the same. Garmin does not make its calculation algorithm public but it is probably based on satellite geometry and other factors such as the URE (user range error) transmitted from the satellite itself indicating its contribution to errors. Many users have estimated that this number represents a 1 Sigma accuracy number or perhaps even a lower 50% number. If you double the number you can be fairly assured that it bounds the real error the vast majority of the time. If you have a differential fix then this error number is likely to decrease even more. On units that are capable of calculating an overdetermined solution (all of the 12 channel parallel receivers), this fact will also be factored in. Don't trust this number as an accurate representation of you current error, it is only an estimate. Older multiplex units and some of the latest firmware releases may be using a more conservative calculation. The way this number is calculated has been changed over the years so if you don't have the latest version of the software you unit may have a different number.

Along the left edge, on many units, is a battery gauge. This graphic represents the battery condition based on alkaline batteries except the gauge on the III family is configurable. While the gauge looks linear the actual performance of a battery voltage is not 6 volt units the 3/4 setting is approximately 5 volts on alkaline batteries which means that ni-cads will not quite reach that mark will fully charged. The gauge will disappear on most units when hooked to external power. A few will show full power under this condition. This gauge in on the main menu screen on other units such as the etrex and emap. On external power these units change the symbol to a power cord. The gauge on 3 volt units is even less accurate that the one on the 6 Volt units since it is quite a bit smaller and more difficult to estimate. A battery will recover itself temporarily if the unit is off for a while so the battery voltage is not a good indicator of battery condition when the unit is first turned on.

The main portion of the screen shows the current almanac from the unit database in a graphical form (on the etrex advanced screen). Each satellite is shown with its number in its approximate position in the sky with the center representing directly overhead and the first circle showing approximately 45 degrees to the horizon. The horizon itself is shown as the second circle. While acquiring a position fix the satellites are shown with North at the top of the display. On some units North will always be at the top even after a fix is obtained while other units will show North as represented by the display choice made on the map page (covered later) or available on the status page menu. The G-III family, emap, newest etrex models and the 76 series permits individual control over this feature from the local menu. The N shown on the display indicates the direction of North. As satellites are received the number in the display will reverse from its starting display. On most units the display shows the number in reverse video and this switches to normal when the unit starts receiving data from the satellite.

The basic etrex shows a simplified drawing with only symbolic satellites shown on the normal screen. (An advanced screen showing entries similar to those above is available from the local menu.) There are only 4 shown which represents the minimum number required to compute a 3D solution. Little visual radio waves are shown when the unit receives a signal from a satellite and 3 such radio waves from different satellites are needed for a 2D solution while 4 are needed for 3D. The unit can track up to 12 satellites simultaneously but there is no indication as to how many beyond 4 it is actually using. There is only one satellite bar shown at the bottom, which will go to full width when a 3D fix is obtained.

The bottom of the page shows a bar chart that indicates the signal strength for each satellite as soon as the unit can receive data from it. The numbers along the bottom of the screen are the satellite numbers as indicated by the internal almanac. On multiplex units there is room for 8 satellites while the newer units have room for 12 indicating the total number of satellites that unit can track. On many units the bars on the chart will initially be hollow and will fill in solid as acquisition proceeds. If your unit starts out hollow then the following interpretation can be applied.

- G-III family, the emap, advanced etrex screen, and G-II+ units - As soon as the unit collects the ephemeris data for that particular satellite the bar will turn solid. Any satellite with current ephemeris data loaded in the unit will be shown as a solid bar. Older multiplex units that begin with hollow bars will also use this interpretation.
- G-12 family - The solid bars indicate which satellites are currently in use for a solution. Therefore no bars will turn solid until a solution has been computed and bars can turn hollow again later if the satellite they represent is not

being used as part of the solution. If you are using differential correction then only satellites that have differential correction applied will be shown with solid bars.

For older multiplex receivers there will be a D shown just under the bars for any satellite using differential correction. This is also true for etrex and 76 family units whether the correction is from a DGPS beacon receiver or the built-in WAAS capability.

Just above the bars and to the left there may be a small icon representing the lamp. (the lamp indicator is on the main menu for the etrex and emap) If the internal lamp is turned on then this icon will be present or if present all of the time it will change shade. Note that this indication will continue even if the lamp itself times out. Pressing any key will turn the lamp back on so you will need to press the key a second time to perform the actual function.

On the G-III family there is also a number that will show up on the right side of the screen. This number is the DOP (Horizontal Dilution of Precision). DOP is a unitless number that indicates the quality of the current fix based solely on satellite geometry. For 4 satellites a DOP of 1.0 would be a perfect geometry while anything below 2.0 would be an excellent geometry. For units performing an overdetermined solution it is possible to see entries below 1.0. It may be possible to see this value on some other units as well using secret start-up commands (see the chapter devoted to this topic).

The etrex Legend, Venture, and Vista have a status (Satellite) screen that varies in several important ways from some of the screens discussed so far. First of all it mimics the emap in not having a position screen but the small size does not permit nearly as much data so this etrex model depends on other screens to fill in the gap. Only the location is added to this screen from the old position screens. It presents a status box at the top and can optionally be shown on other screens as well.

The outer circle can serve as a crude compass by either setting the choices to track up on the local menu or by noticing the small circle on the outer ring present on later units that shows the direction of movement. This feature is also in the emap and 76 units.

These new etrexes, the 76 family, and the GPS-V can also support WAAS. If WAAS is enabled then the upper two satellite positions are reserved for WAAS satellites leaving 10 satellites for normal navigation. The standard set of satellites will always have a number between one and 32 so any number above that can be used to identify the WAAS satellites. In the illustration shown in the DPGS chapter these are 35 and 47. Note that these are also shown on the screen in their expected positions on the sky map. The number 35 is shown ESE and 47 is shown WSW from the location displayed. Note that until the satellites have been identified the first time they will be shown in the north position but once located and identified they will be shown in their true positions. They will provide differential correction data and may be used as satellites in the position solution as well. If they are present the differential corrections will be indicated with small "D"s on the satellite bars.

The 76 and Map76 have much in common with the emap and new etrex display. They can support WAAS but because they have the larger screen size like the emap they can also support more data and include three user customizable fields. They, too, do not have a position page.

Function Keys for the Status Screen

The arrow keys can be used to set the screen contrast. Press enter to accept the change or quit to cancel. Note that the emap has a separate key that can be depressed and held down for a second to bring up a contrast menu at any time. Similarly the 12CX uses the map button to do this on any screen. Therefore for the lamp function to actually change the lamp you must press the button a second time. The etrex uses its up/down key for this.

The local menu can be used to aid in acquisition. Press enter on most units or the menu key once on the III family (or emap). The menu will show the following choices:

- Start Simulator - III family only. All units can set this from the system menu.
- The new etrex units have a similar, GPS off, function and a demo mode (from the main menu) that will cause the speed to be set to 20 mph.

- Track Up - III family, emap, etrex, 76. Others set this entry on the map screen. This sets the orientation for the satellite display. If you already have set Track Up then then the option will be 'North up'.
- AutoLocate - See the chapter on obtaining a fix to see how this works.
- Initialize Position - receivers with maps only (called new location on the emap) - Initialize position by selecting a position on a map. You only need to be within a few hundred miles. You can use the arrow keys to pan the map and the zoom keys to move in or out. Press enter when you have reached the proper location. On the emap a second choice is presented called automatic. This choice performs an autolocate.
- Select country - receivers without maps - initialize position from a list of countries or states. Use the arrow keys to move down or up the list. If you hit the bottom or top of the page it will move a full page at a time. The states are listed under the prefix US.
- Set 2D altitude - III family, emap, etrex, 76. On others this is set on the position page to the approximate altitude to aid in rapid acquisition. Setting this will also provide more accuracy if you only have a 2D fix and can aid in acquisition if the altitude has changed significantly since the last time the unit was on.
- Continue acquiring - Not on the III or emap family. Tells the unit not to prompt you to initialize it.
- Show GPS elevation - Only on the Vista to display the GPS calculated elevation. Normally the Vista uses the built in altimeter as the source for elevation data.

Note that the basic etrex does not offer an initialization option, but figures out the best solution on its own and will attempt to use your local location. If this fails it will do an autolocate. You cannot seed this unit with an altitude to help it with obtaining a fix. The lack of these features may make it difficult to achieve a rapid fix in an airplane or after you have moved several hundred miles but will not generally cause any problems.

Hitting the page or quit key will rotate to the next or previous screen in the sequence. If you haven't done this while awaiting the fix then the unit will automatically switch to the position page as soon as it has a 2D fix. (G-III screen illustrated.) Some of the latest units are missing this page.

The Position Screen

Figure 9 Position Screen

The position screen focuses on telling you information about what is happening at the current time. It provides such data as current speed, current distance, current position, current altitude, etc. For all units except the mapping units it looks very much like the G-12 screen shown second from the left on the top row in figure 7. The III family above in landscape mode looks similar except that there are 6 pieces of information in the center area instead of 4. Note that the GPS V calls this screen the trip computer.

Since much of the position data is repeated on the navigation pages the etrex has taken the approach to eliminate this page entirely and display more selectable data on the navigation page (see the navigation chapter for more details) or on the trip computer page. The emap displays position data on the GPS info page that combines it with the status screen. This works since the emap has a larger display than other handheld units. The emap shows the current track direction on the outside ring of the satellite display. If the satellite display is set to track up then the whole display rotates to keep the track direction at the top. If the display is set to North up then a small circle appears on the outside ring to display the direction of movement. This small circle moves around the ring to indicate that direction. Both the etrex and emap move most of the trip data to a separate menu page.

Screen Contents

This screen is dominated by the edge view of a compass. While there is no compass in your GPS, the unit can simulate a compass in software whenever you are moving. Using computations based on previous position solutions and Doppler shift data it can deduce both your speed and direction, i.e. it can compute your velocity. The compass display shows the direction of your movement, called a track, in a graphical form. The next entries on the screen depend on which model you own.

- Older Multiplex receivers will display the 4 entries shown in figure 7. These are Track, Speed, Trip, and Altitude where Track is the numeric form of the compass display, speed is the calculated current speed in the units of your choice, trip is a resettable odometer, and altitude is the current computed altitude. It is settable if you have a 2D fix. If you don't have a fix at all then the bearing and speed fields will be nulled out.

- Non-mapping 12 channel parallel receivers in the G-12 family and the II+ will show exactly the same 4 entries as multiplex receivers in their original default configuration. However both the Trip and Alt entries can be changed to display different data if you prefer. The Trip entry can be changed to display TRIP (trip odometer), TTIME (Trip timer), ELPSD (Elapsed Time), AVSPD (Average speed traveled), and MXSPD (Maximum speed traveled). The ALT entry can be set to display ALT (Altitude), TRIP, ELPSD, or TTIME.

- Mapping receivers in the III family have 6 customizable entries. In addition the G-III family can independently change the entries for vertical or horizontal orientation. The choices include: Altitude, Average Speed, Battery Timer, Max Speed, Odometer, Speed, Sunrise at present position, Sunset at present position, Track, Trip Odometer, Trip Timer, User Timer, and Voltage.

- The emap shows speed, altitude (called elevation), and the satellite display can serve as a crude compass like display of your current track. The actual track as a numeric readout is not displayed anywhere on the emap. Since the emap is not intended for marine or aviation applications displaying your actual travel direction with a high degree of precision is not needed.

The speed entry on this page displays only the horizontal speed component and some have wondered about the accuracy on slopes. Certainly it can be inaccurate in a hot air balloon or perhaps the steepest ski slopes but generally this design decision results in very little degradation of accuracy. On roads and trails a 10% slope would be considered very steep and even this much slope would cause only a 0.5% difference in the speed. Others have wondered about the accuracy of a GPS device in general to record the speed that you are traveling. At road speeds the accuracy is better than most car speedometers. It is typically within 1/2 mph of the true speed for steady travel.

Three entries on the position page are tied very closely together. These are trip distance, trip time, and average speed. Resetting any one of these will also reset the others. There is some debate on exactly how this information should be calculated and on the G-III family you can specify which of two methods you would prefer. When most folks think of an average speed they simply expect that you should divide the total elapsed time of the trip into the distance for the trip. This particular averaging technique is simple and correct when you have no control over the stops that may happen during the trip. However, some times you do have control over the stops, especially on a hike. In this case you might prefer the average speed to indicate the speed that you were actually traveling when moving. In this way you can estimate how long it might take to return to your starting point if you didn't take any rest stops. This is the averaging method used by Garmin receivers and the trip timer does not increment whenever you are stopped. In addition there is no calculation of distance made when the unit is turned off or during periods of poor coverage.

The emap, legend, venture, vista, and 76 series keeps this data on a separate page called the trip computer that can be reached from the main menu. It computes both types of averages simultaneously. It also displays max speed on this page along with a second odometer. A local menu item is used to reset the trip meter, odometer, and max speed. The basic etrex displays this data on the navigation (pointer) page that has a local menu to reset it.

Near the bottom of the screen (top on the emap) you will find the actual position indication. This will be displayed in the grid system and datum of your choice. When you don't have a fix or in simulation mode you can select and modify this data except on the mapping receivers. Changing this information is one way to enter waypoints or may be used to help the receiver achieve a rapid lock. To change the grid system or datum you will need to go to the main menu (shown in figure 7 above at the left side in the middle position) under setup/navigation setup.

Beneath or beside the position data you will find a display of the current time. The time zone being displayed is a user specifiable option but you cannot change the time itself. This is computed as part of the position fix and is then corrected for leap second data and displayed on this page. Internally a GPS computes the time to accuracy in the nanosecond range but the update of the display is a low priority task thus you can generally expect the display to be within a second or so of the actual time. Changing the time zone is accomplished from the main menu by selecting Setup/System setup. You will need to offset the timezone setting for daylight savings time. The etrex and emap display the local time on other pages and can automatically switch from standard time to daylight savings time if you wish.

Function keys for the Position Screen

There is only a local menu available for the G-III family receivers. It allows you to average your position (and then save it as a new waypoint), change the fields at one of the 6 customizable locations, or restore the factory defaults.

- To average your position you should highlight the selection using the arrow keys and then press enter. Select Save when you have completed your averaging or Discard to cancel. If you select Save and press enter you will get the standard waypoint entry screen to name your new waypoint if you wish to change the default name.

- To change a customizable location you should select this option and then use the arrow keys to select the entry you wish to change and press enter. You will then get a menu that you can scroll to determine which item you want to display at that location. Press enter to perform the selection.
- You can reset the 6 entries to the factory default by selecting the factory default option.
- Other Non mapping Garmin units use the direct object oriented paradigm. Use the arrow keys to highlight the object on the screen and then hit the enter key to select the object you wish to change. Anything that can be highlighted can also be changed. This will vary depending on the state of the unit at the time you press the keys. For example you can select the altitude data only if you currently do not have a 3D fix. This can be used as an indicator for a 2D position. If the ALT entry is being displayed on the screen you can use the up arrow key to select the altitude data rapidly. If the data in the altitude entry highlights then you do not have a 3D fix. If you leave this highlighted then it will automatically become unhighlighted the moment a 3D fix is obtained.

If a data field is highlighted it can be changed by hitting the enter key. Many of the fields can only be reset and the word 'reset?' will appear allowing you to use the quit key to change your mind. Press enter to perform the reset. On entries that let you change the data itself you can use the left/right arrow keys (left arrow at the beginning to clear the field if desired) to move to the number or letter you wish to change and the up/down arrows to change it. Continue until all of the desired changes are made and then press enter to complete the change. Note that on multiplex units the value in the trip meter can be changed but on newer units it can only be reset to zero. On the G-III family the values can be reset by using the main menu (the status screen local menu for altitude).

If you highlight the text above the data then you can choose which data is to be displayed by hitting the enter key. You can use the arrow keys to scroll through the choices available. The current data for that choice will also be displayed so you can use this feature to review the contents of the entries without actually selecting one for permanent display. Hit the quit key when you are finished or the enter key to select that entry for permanent display.

The Map Screen

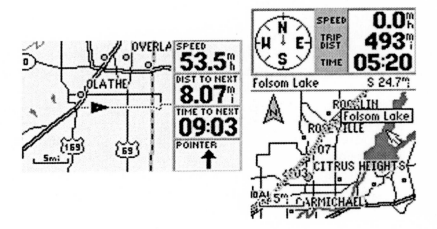

Figure 10 Map Screen

Pressing the page key while on the position screen will take you to the map screen (G-III version shown in figure 10 to the left). On units without maps this might better be called the plot screen. It contains a graphical representation of your current position with respect to nearby waypoints and perhaps your tracklog. A sample for the G-12 is shown in figure 7 above as the third screen from the left on the top row. Mapping receivers will also show an underlying map on this screen although this feature can be turned off on all except the emap and etrex models. Some receivers with a city database may also display cities on the map. For purposes of this discussion cities, if present, will be treated like a displayed waypoint. Please see the chapter on databases for information on maps and the city database. The map screen is both a visual orientation screen and a screen for navigation use.

Screen Contents

What you see when this page appears depends a great deal on which model you have. Everyone will see a symbol in the center or just below center of the display that represents his or her current position. In addition you may see up to four pieces of text data either at the corners of the graphic or in a row at one edge. Depending on the current zoom setting and what waypoints you have stored you may see some of those as well.

On the mapping receivers you will likely see a map in the background. You can turn off the 4 text entries and make the map fill the whole display. The four text entries on the III family are customizable just as the entries were on the position page. The GPS-V has 4 customizable entries similar to the III family but the display automatically changes when navigation is in progress (see the "Autoroute Chapter" for details). Since the III family and V have rotatable displays a different set of 4 entries can be programmed into the rotated display. The 4 entries on the emap are not customizable but will change automatically when navigation is in progress. The four entries on the emap, shown above in figure 10, include a small compass display to indicate the track, a speed, a trip distance, and a current time display. The symbol in the center on these units will turn as you do so that it always indicates your present course. On the III Pilot it looks like a small airplane while the others display an arrowhead. The etrex mapping receivers behave similarly but only contain two customizable data displays along with the map.

On older non-mapping receivers you will see up to 4 numbers in the corners of the display. These are the main four significant pieces of data and are available on this and both navigation screens. These always include your track (direction over the ground) in the lower left corner and speed in the lower right. If you are actually navigating to a waypoint then the upper left corner will indicate the bearing to that waypoint and the upper right corner will indicate the distance to that waypoint. Speed and track will be reset to null values if you lose a fix.

Some units will display a banner across the top of the screen. Units that do this include the new etrex units and units that do not have dedicated zoom keys. The menu choices include a zoom selection, a pan selection, and perhaps an option or config selection.

The etrex click stick units include the ability to show the current fix status on the map page.

There is usually some sort of scale indication on the screen. Mapping receivers have a map like legend for scale and perhaps other legend data showing that you have overzoomed (the scale of the display is too high for the level of accuracy of the map, not the GPS) or whether you are using loadable maps. Non-mapping receivers will show the scale by indicating the full screen height in the zoom setting or will display rings with a legend on the rings indicating the distance in the units of your choice at that point. You may find that the display is indicating .4 miles and you really want to know how many feet that is. For English measure a rough rule of thumb is that, since there is about 5000 feet in a mile, then a ring distance in thousands of feet is equal to the zoom scale. So for the above case the zoom of 2 miles shows a ring at 2000 feet. Metric rules of thumb are left as exercises for the user.

Function keys for the Map Screen

On units with a banner at the top something will always be highlighted. If you hit the keypad or click stick keys the item highlighted will change. Hitting enter will select the highlighted object. Selecting the zoom function will permit using the up/down arrows to change the zoom factor for the display. Press enter to accept the change or quit to cancel. Selecting pan will permit the arrows keys to be used to pan the screen. On these units you should notice the zoom button is highlighted during the pan. Hitting enter will permit zooming in and out while panning.

Receivers with zoom keys can use these keys at any time to change the map's screen scale. The arrow keys on these machines are used to perform panning operations. Use the rocker keypad to move the cursor and as you hit the edge of the screen it will automatically pan the data beneath the cursor. An arrow indicates the current panned location and object information will appear if the arrow is positioned over a map object. Folsom lake is shown on the emap figure above as identified with the arrow. The classic etrex and summit uses its up/down keys for zooming and has no pan function.

Etrex and mapping receiver tip - While the classic etrex does not have any panning keys you can usually pan easily to any point of interest. Just select a waypoint and then use the show map command to jump to the area around that waypoint. The up/down keys can be used to zoom around in the panned area. This trick will also work with mapping receivers which may be easier and quicker than trying to pan to the location.

Some units will actually pan in 8 directions by pressing the rocker keypad on a diagonal. If you happen to pan over an object you can select that object with the enter key just as in normal mode. While panning the bearing and distance numbers will report distance from your current position to wherever the cursor is pointing. Some units will also report the lat/lon position. Press the quit (esc on the emap, page on the etrex) key to leave panning mode and return the screen to the current position. Use the pan function to review the tracklog the active route or look for cities. It is not particularly useful as a tool to look for waypoints since only the 9 closest to your current position will be shown, limited to a distance of 100 miles (82 on the emap). The emap and etrex, legend, venture, and vista show the 15 closest waypoints and when panned it will show the 15 closest to the panned location rather than the current location.

On the III+, 12map, emap, vista, legend, and map76 you can also highlight a freeway exit while panning. Pressing enter will bring up an exit information screen that tells about services available at the exit. Other features related to services are covered in the chapter on databases.

Some receivers have a special local menu shown in the upper right of the screen. If they have a banner menu and the word cfg or opt is listed in the banner then you can select this to bring up the menu. On units with zoom keys and no banner use the standard local menu key (menu or enter with nothing selected) to bring up the menu. If your unit doesn't have a local menu then these settings are on the main menu screen.

G-III family local menu settings include:

- Data Fields OFF/ON - turn off the data fields allowing the map to occupy the full screen.
- Change Fields - customize the 4 data fields. You can choose such things as Average speed, Bearing, Distance, ETA to Destination, Speed, Time to Destination, Track and trip odometers, a graphic arrow pointing to the next waypoint, and other navigation settings. (See the chapter

on navigation for information on what these fields mean and how to use them.)

1. Select Change Field from the local menu with enter.
2. Highlight and select the field you wish to change.
3. Scroll to the desired field data and select with enter.

- Measure Distances - click enter on the point you wish to measure from and then move the cursor using the arrow keys. The bearing and distance will be displayed at the top of the screen.
- Restore Defaults - reset to factory settings.
- Map Source Info - III+ only - Displays information about downloaded maps. You can see what is downloaded and how much memory it consumes out of your 1.4 Meg. You can also choose which map will be selected for viewing. This is primarily used to force one map when two cover the same area. To select a map highlight the box to the left of the name and use the enter key to toggle an X in the box.
- Setup map - configure map preferences. The entry will show tabs across the top which can be selected to perform the following selections:
  - Map - Orientation, AutoZoom, Land Data
    - Detail - G-III+, This setting applies to any map feature that is set to automatic. It specifies the scale at which the feature appears. If you have set a maximum or off somewhere else then that setting will override this one. Settings are more, normal, less.
    - Orientation - Course Up, North Up, Track Up. Sets the map direction on the screen.
    - Automatic Zoom - on/off to change the zoom when nearing a destination waypoint. Automatic Zoom ranges from a maximum of 80 miles to a minimum of 800 feet. This zoom mode will attempt to keep the active route waypoint in view at all times. It will zoom in or zoom out as necessary to provide maximum detail while showing the waypoint. If the user manually zooms the screen then the unit

will honor the user selected zoom setting until the automatic zoom would move beyond this setting in the direction the user chose. If the user zooms such that the automatic setting does not reach the user selected zoom then the zoom setting will be reset as the unit passes the active waypoint and selects a new one. For example if the user zooms in closer that 800 feet or zooms out while the unit would normally be zooming in. Note that at higher speeds the unit may never actually zoom in as far as 800 feet. Generally the automatic zoom only zooms in since the user is usually making progress toward the waypoint.

- Acc. Circle (Accuracy Circle) - This is a circle of uncertainty with regard to you current fix and its relation to the underlying maps. It takes into consideration the current EPE calculation, but mostly it considers the accuracy of the underlying map. To minimize the size of the database the underlying maps are stored with a certain amount of precision. This is why you often see "overzoomed" as a message on the map screen. The accuracy circle is another way to display this same information. If you turn the map display off you will see increased accuracy for this circle, on the III+ if you load a MapSource map for the area you will see increased accuracy, and also if you are using a DGPS you will see a smaller circle. Note that is on the line tab on the G-III.
- Land Data - G-III to turn the display of land data on or off. G-III Plus has this on the Source tab.
- Source - G-III+ - selects which of the available maps you wish to turn on. Since this unit has loadable maps then this feature can turn off these maps to view the base map in overlapping areas.
- Line - Track log, Active Route Lines, Lat/Lon Grid, Acc. Circle

- Track Log - Log of your current movements. Display can be turned on or off with this setting.
- Active Route Lines - can be turned off or on.
- Lat/Lon Grid - G-III can be turned off or on. Helpful to get an understanding of the scale.
- Acc. Circle. - G-III has this here. See discussion on the Map above.
- Roads/Road labels - G-III+ has this setting here.
- Wpt (called Pts on the III+) - Waypoints, waypoint text, Active route waypoints - These items can be turned off or set to appear based on a zoom level. Only the nearest 9 waypoints will be shown when this feature is turned on. Text font size can be specified. III+ has exits setting here, which can display exits for major freeways.
- City - Large/medium/small cities and text - Cities between 5,000 population and 100,000 are considered medium cities. You can turn off the display of these items or set them to appear based on a zoom level. Really large cities will always be displayed. Text font size can be specified.
- Road - G-III - Freeway, National Highways, Local Highways, Local Roads - These can be turned off or set to appear based on a zoom level. - G-III setting, the G-III+ has this under line.
- Other - State/Prov, Rivers/Lakes, Metro Areas, Railroads - These items can be turned off or set to appear based on a zoom level.

The emap has similar features to the III family. The interface has been simplified somewhat. The accuracy circle is always on but otherwise works as described above, you cannot turn off the base map, and it has less options in the display of data. It has the following items on the local menu:

- Use Indoors/Use Outdoors - Turn on simulation mode or return to normal mode.
- Full Screen Map/Show data fields - toggles between a full screen map and a map with data fields at the top.

- Show Next Street - This feature will only work to show the next street if a MetroGuide map is installed in the optional cartridge. However, it operates as a next exit for the freeway exits in the basemap. It displays the upcoming cross street as you drive or upcoming exits, which are shown with a distance to the exit as measured to the overpass, not to the exit ramp.
- Start/Stop Navigation - If you are navigating using a trackback, a route, or a goto you can suspend and restart the navigation by toggling this menu item. This is useful to change the display at the top of the screen or to force the unit to recalculate its display of average speed or recompute a starting point in a route.
- Measure Distance - will show a distance from the current position or any position after you hit the enter key.
- Setup Map - This controls the display of information of the screen. You have three tabs with settings to make.
    - Map - Controls the detail of map features, map orientation (North up or track up), and <u>autozoom</u> as described above for the III family.
    - Line - Controls the tracklog display (on/off) and settings (see the tracklog chapter for details.) as well as street labeling. Note that the tracklog on the emap is always on and always set to automatic, it is only the display that can be controlled here.
    - Road - Controls lock on roads (only works with MetroGuide, or City Select maps). This permits the unit to force the GPS display to lock to an existing display. Street Label text size and zoom control is also on this menu.
    - Other - controls text size and detail for POI's, zoom level, Land cover.

The etrex mapping receivers have features very similar to those shown above:

- Pan Map - Allows map panning. Objects can be selected and distances are shown from your current location. In addition a readout displays the lat/lon location of the panned position. Waypoints can also be set while in this mode

- Stop Navigation - terminates navigation (does not just suspend like the emap does.) If you think you want to resume a goto then it would be a good idea to save the location in your favorites list before canceling the goto.
- Hide Nav Status/Show Nav Status - Hide or show the status window that was available on the satellite status screen.
- Hide Data Fields/Show Data Fields - Toggle full screen map mode.
- Setup Map - Displays map setup screen
  - Page
    - Orientation - Track Up or North up
    - Auto Zoom - on/off. See description on III family
    - Detail: Most, More, Normal, Less and Least
    - Lock on Road: On or Off - only works with MetroGuide or City Select Data
  - Tracks
    - Saved Tracks - Scale options, off, Auto or 120 ft - 800 miles.
    - Main Tracklog Scale options
    - Goto Line: Bearing or Course, Bearing is constantly updated from your current position while Course is set at the start of Navigation and thus shows crosstrack error visually.
  - Map Features
    - Points of Interest - Scale options
    - Waypoints - Scale options
    - Street Label - Scale options
    - Land Cover - Scale options (This is the land, lake outlines, etc.)
  - Text
    - Points of Interest - Sizes from None, Small, Medium, or Large
    - Waypoints - Sizes
    - Street Labels - Sizes

- Land Cover = Sizes
- Transferred MapSource Map Data (Only if Maps have been download) - Lists all downloaded maps and permits a toggle of on/off to display the map. The order is determined by the order they are listed in Mapsource.
- Restore Defaults - Return map page to the factory defaults

On non-mapping units the local menu will permit changing the map or tracklog settings. Tracklog settings are covered in the chapter specifically devoted to tracklogs. The map settings entry selects map preferences. You can select from the following items:

- Track Up/DTK UP/North up - This setting will determine how the display will be oriented. You can have the top of the display always point north (North up), always point in the direction your are currently traveling (Track Up), or on some units you can have Desired Track Up which means the current route leg or goto direction will be up. Note that this setting may also effect the status screen display. If you have dtk up or track up set then the current position location will move downward to permit more data ahead of the current position to be shown more clearly. The etrex does not have DTK up.
- Rings - Turns on or off the display of the distance rings. These are helpful to estimate distances on the display. They have 3 rings equally spaced although you may only be able to see the corners of the outer ring. (not available on the etrex.)
- Routes - Turns on or off the display of routes. On the etrex active routes are always shown but you can suspend navigation to turn them off. See below.
- Nearest - Turns on or off the display of the 9 nearest waypoints. (On the etrex all the waypoints are shown always.) Note that it is not possible to show more than 9 waypoints or points further than 100 miles (250 miles on the etrex) unless they are part of an active route.

- Names - Turn on or off the display of names. Icons will always be shown. If you have names off you can select the object to view its name. The etrex always shows names.
- Track Log - Turn on or off the display of the tracklog. On some units older units you can specify the length of the display of the tracklog. The etrex always shows the tracklog.
- Start/Stop Navigation - etrex only. You can suspend an active navigation using this menu item to temporarily turn off the display of the route.
- Auto zoom on/off - You can automatically have the unit zoom in when you get close to a waypoint in a route, see the discussion of this feature above under mapping receivers.

You can also set waypoints using the map display. If you are panned to a spot on the screen where nothing is selected and hit 'Mark' you can set a waypoint directly at that point in the display. This is useful for visually averaging several tracklogs or setting your own route based on a tracklog or underlying map display. You can also navigate directly to a location on the display on units with a goto key. Pressing the 'GOTO' key will build a waypoint named 'MAP'. If you hit 'ENTER' at this point your GPS will automatically start navigating to the MAP waypoint. Rename this waypoint if you wish to keep it since it will be overwritten by the next use of the map waypoint feature. If a waypoint is already highlighted then pressing 'goto' will select this point as a target for navigation and pressing enter will select the waypoint for viewing or modification.

Data Customization

In addition to the screen display customization that can be set using the map setup menu there is often data display customization as well. If you are able to highlight the title of a data field then it can be selected and changed to display the data you wish. For example the III family can use any of the four fields to display a number of kinds of data including current information, information about the unit itself, and navigation data. These include Altitude, Average Speed, Bearing,

Course, Distance, Distance to Destination, ETA, ETD, Fuel, ETE, Max Speed, Odometer, Off course, Speed, Sunrise at present position, Sunset at present position, Track, Trip Odometer, Trip Timer, Time to Go, direction pointer, and more. For the mapping etrex units you can select two of: Bearing, Course, Current Destination, Current Distance, Current ETA, Current ETE, Elevation, Final Destination, Final Distance, Final ETA, Final ETE, GPS Accuracy, Glide Ratio, Glide Ratio Destination, Heading, Maximum Speed, Moving Average Speed, Odometer, Off Course, Overall Average Speed, Pointer, Speed, Sunrise, Sunset, Time of Day, To Course, Trip Odometer, Trip Time – Moving, Trip Time, Stopped, Trip Time – Total, Turn, VMG, Vertical Speed, and Vertical Speed Destination. The 76 and Map76 are also fully customizable with similar entries to the list above and can also determine the number of entries you wish to display on the screen, 3, 6 or 9.

More information on using the map screen in navigation and a definition of the terms used above will be found in the navigation chapter.

Other Screens

The next screens in the rotation are the navigation screens, which are shown above in figure 7 as the two right hand entries in the top row. These are called the compass screen and the highway screen. They may be reached by successively hitting the page key or you may have only one of them in the rotation. If you have only one in the rotation then the local menu for that screen can be used to select the other one. You can customize which screen appears in the display rotation. An explanation of these screens and how to use them is in the navigation chapter. The etrex has only one navigation screen (called a pointer screen) while the newer etrex units have this screen as well with two customizable entries similar to the choices on the map screen and they have the ability to display navigation data on the trip computer screen. The emap does not have a dedicated navigation screen at all, but modifies the entries on the map screen for navigation use.

Other screens that you may find in your screen rotation include the route screen, a trip computer screen, and perhaps the main menu.

The route screen is described in the routing chapter. The main menu is described as needed in other places in this manual. The main menu has many submenus that are shown in figure 7 above. The main menu and its submenus are traversed like any other screen. You use the arrow keys to highlight the item you wish to view or change and then hit the enter key to select it. Screens also appear when you press the mark key or the goto key. These are covered elsewhere in waypoints and routes. Units having a Find key will bring up screens associated with waypoints and database objects. Pressing and holding the find key will switch to the route screen if a route is active or the waypoint screen showing the destination if a goto is active.

The GPS-V has some screens that are unique to autorouting and autoroute navigation. These are covered in the Autorouting chapter.

Trip Computer

The trip computer screen is a feature of the newer Garmin receivers. The G-V trip computer screen is the same as the position screen on other units. The etrex Legend, Vista, and Venture have a trip computer screen that is customizable and can substitute for an additional navigation screen that presents only text data, or a position screen. It even has a choice of large numbers with only 4 selectable fields or 8 selectable fields of differing sizes. The choices available for display on this screen include the ability to show position data in two different grid systems simultaneously. One choice is always Lat/Lon while the other choice is the same as the current grid choice. The choices for display include: bearing, Course, Current Destination, Current Distance, Current ETA, Current ETE, Elevation (altitude), Final Destination, Final Distance, Final ETA, Final ETE, GPS Accuracy, Glide Ratio, Glide Ratio Destination, Heading, Location (lat/lon), Location (selected), Maximum Speed, Moving Avg.

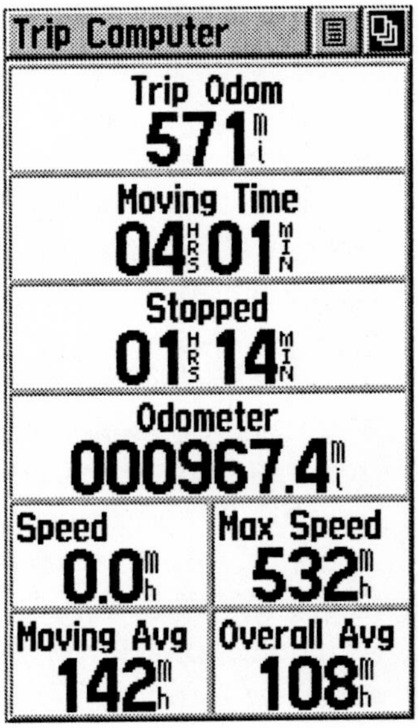

Figure 11 Trip computer

75

Speed, Odometer, Off Course, Overall Avg. Speed, Pointer, Speed, Sunrise, Sunset, Time of Day, To Course, Trip Odometer, Trip Time - Stopped, Trip Time - Total, Turn, Velocity Made Good, Vertical Speed (Vista only), and Vertical Speed Destination. For definitions of the navigation terms included in the above list check the navigation chapter. Not all of these choices are available in the narrow fields.

Other units with a trip computer, such as the emap and 76 family, do not have a customizable display but provide an odometer, a trip meter, maximum speed, and other trip related data. Trip average speed is calculated by providing two solutions, one that considers stop time and one that only considers moving time. Trip time itself is also shown both ways. Showing data in this fashion is very useful if you intend to use the trip computer to help you decide how long it will take you to get back to your starting point. If you have no control over the stop times, such as ones caused by stoplights, then the time including stop time is the one to use. But, if you have control over stop times such as when you are on a hike and only stop when you want to rest or study something then the moving time shows you how long it would take to get back if you didn't want to stop.

Altitude Profiling

The etrex summit, etrex vista, and 76S also have an altitude profiling screen. This is a special screen in the normal page rotation that is devoted to vertical navigation features. This screen will provide current elevation and an elevation profile similar the tracklog on the map page except that it is devoted to vertical movement. It will also display your vertical speed (ascent or descent). Selecting one of the two (on the vista) user selectable fields using the up/down arrow keys will permit choosing one entry from the following list: Local Pressure (renamed to Ambient pressure on new Vistas), Max Descent, Max Ascent, Average Descent, Average Ascent, Total Descent, Total Ascent, Min. Elevation, Max Elevation, (12 hour pressure trend for Summit), and for Vista vertical speed and normalized pressure (renamed to Barometric pressure on new Vistas). The 12 hour pressure trend can be used to help predict the weather. Normalized pressure is adjusted to show what the pressure would be if the altitude was zero (Sea level). The use of the altimeter and this screen is described in the miscellaneous chapter covering product unique features.

Chapter 5
Working with Coordinates and Units

This chapter covers the various grids, datums, and other measuring items in a Garmin unit. Much of this information is mathematical but it is simplified wherever possible just to provide an overview of the information. This simplification could lead to errors or misunderstandings of details so please be careful when drawing conclusions from this explanation.

The datums and grid preferences can be reached from the main menu by selecting the navigation setup options. You can set the navigation datum, the navigation grid, the user units, and if your unit support cdi, this can also be set here.

Datums

What is a datum? First, what it is not. It is not a singular form of the word data which some of you may have heard. Data is a collective noun and the plural of datum is datums when used in a geophysical sense. For GPS, navigation, and geophysical use a datum represents a reference point from which you can measure things. At one time this was a location on the earth and could have actually been just a single point of reference. In modern usage it defines not only the point for reference, it also defines the model of the earth used to support those measurements. The model is a three dimensional one that describes mathematically the shape of the earth, the location of the center, and the location on the surface that represents the starting point for measuring. Some older datums only defined a horizontal measuring model or, in some cases, only a vertical one. Some datums permit measurements to be made worldwide while others are only defined to support a local grid measuring system. If you were to use a different datum to measure the same place you would typically get different results and is some cases drastically different results. Internally Garmin units always store information and compute using the WGS-84 datum but will display the answers in the datum of your choice.

Typical Garmin handhelds let you define any of over 100 datums for use anywhere in the world but there is no check that you used an appropriate datum for the area you are currently located in. The question is: "Why would you want to define a datum?" Since a GPS unit makes all measurements using the WGS-84 datum this might seem enough to just display the answer using that datum. In fact this may be enough for some folks and some GPS units only support this one datum. However, surveying and map making have been going on for centuries. If you have a map created before the modern age of GPS your map probably was not made using the WGS-84 datum. If you want your position data to agree with the map or survey marker on other source of geophysical location information you will need to ensure that your GPS datum agrees with the one used to generate the data. Generally, paper maps provide this information in the legend of the map itself. In addition some maps use a local grid system (covered later) that requires the use of a specific datum to maintain the accuracy of the grid system.

As mentioned before many datums are only defined for the horizontal plane and thus your GPS will continue to use the WGS-84 datum for vertical (altitude) information. This datum defines zero height using a mathematical model of an ellipsoid, which is basically a ellipse spun around its minor axis to form a globe shape. The surface of this ellipsoid is considered to represent zero altitude with points above the ellipsoid representing positive altitude and points within the model negative altitude. This model does not take into consideration the many real life changes are caused density differences in the earth and earth motion. What we would prefer is a measure of altitude that represents zero with sea level, or more precisely mean sea level. To get this number your GPS uses a table that has been defined for just this purpose. It is called a geoid and can permit the translation of GPS computed altitude to its mean sea level equivalent. It should be pointed out that while we seem to have defined altitude very accurately, a handheld GPS is not a particularly accurate device in measuring altitude. Because of the geometry of the satellites in the sky and the fact that the earth's surface blocks our view of most of the constellation the accuracy of a vertical fix is about 50% worse than the accuracy of a horizontal fix. In addition there can be some errors in the geoid transformation due to the simplified table. Unfortunately this is often disconcerting to users since they generally know their altitude to better accuracy than they know their horizontal location and even may measure it with more precision (feet vs. miles). Do not let this difference cause you to believe the GPS itself is not reporting an accurate fix.

Translations among the various datums can be a cause of some errors. Garmin uses the Molodensky transform parameters for those datums and performs transformations as needed. This is a simplified model and can result in errors on the order of 10 meters in some cases. With this transform the WGS-84 and NAD-83 numbers are always the same while a better translation will show a slight difference in these two datums. For most of the USA this difference is less than a meter rising to about 2 meters on the west coast; still less difference than the accuracy of the unit in most cases. Modeling something as complicated as the earth does not lend itself to mathematical transformations so to achieve accurate results some information is contained in tables with model information used to access the entry in the table which is then interpolated to get the answer. It is not known where Garmin depends on models alone and where they supplement with tables.

WGS-84

WGS-84 is a worldwide datum and is the master datum used by a GPS. Garmin always stores internal information in this datum. The origin is the center of the earth and then an ellipse is defined using the major and minor axis. Information about ellipse flattening and the gravitational constant is also part of the definition. The latter is used to help calculate the geoid height. The model is augmented with stations at precisely defined locations on the earth's surface to pinpoint the accuracy of this system.

As more and more mapping systems have become digital in nature and databases have begun to be the norm for surveying there is a tendency to use the WGS-84 datum more and more. This is exacerbated by a need to share data from around the world that was originally generated on a different datum. Hopefully this trend will continue as it simplifies data sharing and reuse of information.

USA datums

There are several datums that have been or are still being used in the USA. One of the NAD-83 is almost identical to WGS-84 and can be considered for navigation purposes to be the same. The differences are well under a meter most places in the USA and are caused by a slight difference in the constants used in the calculations as developed by the different agencies administering the datums.

Many existing maps use the NAD27 datum. The difference in coordinates between NAD27 and NAD83 can vary up to 10's of meters. The 1927 datum was determined from less accurate and fewer observations. Factors such as seismic motion have changed station positions over time and computational capabilities did not exist to average readings. The newer datums are geocentric while this datum was based on a point in Kansas, and gravity data was not used in the network adjustment of the 1927 datum. There are other NAD27 datums with other locations chosen as their reference points for areas not too close to Kansas. Most USGS maps were generated using the NAD27 CONUS (continental US) datum.

Other Local Datums

There are datums in use all over the world and you need to be careful to use a datum that matches the map you are using. Study the map legend for this information or check with the map manufacturer. Some map grid systems assume a specific datum is used such as the British system used in Great Britain. Very few paper maps use the WGS-84 datum so if you need to relate your position to a paper map you will have to deal with the datum question.

User Defined Datums
The WGS-84 datum relates to other datum in common use around the world in terms of differential X, Y, Z coordinates (DX, DY, DZ) and DA and DF which are the differential equatorial radius and differential flattening. This information can be entered into Garmin units that support user defined datums to provide the datum of your choice.

Grids

In addition to defining the origin and model to use for measurement purposes the GPS receiver with also have a choice of grids to use to display the horizontal location.

Lat/Lon

Everyone has probably heard the terms latitude and longitude as referencing a grid system that covers the earth. This system has been around for hundreds of years but gets a new dressing for GPS use. Latitude and Longitude assume that the earth is a big ball and defined as a spheroid. The fact that it rotates around the poles on each end is used to develop a grid system that is based on this angular motion of rotation. Basically the idea of longitude and latitude is to measure the angles represented if you were slice the earth into a circle. If you slice the earth at the equator and then divide it up you would have a longitude line at each degree around the circle. These lines would all meet at the poles. Rotation of the earth would dictate that the sun would travel the same angular longitudinal distance in the same time. Each one of these lines is called a meridian. If you slice through a meridian you have a circle and each degree around the circle is called a degree of latitude. While longitude lines will all meet at the poles, latitude lines are all parallel to each other.

For more precise measurements the degree is divided into 60 minutes and the minute is divided into 60 seconds just like time. For calculation purposes measurements and calculations are often done in decimal fractions of a degree or decimal fractions of a minute. Seconds are always divided decimally if a smaller unit in needed. Garmin GPS units can be set to display lat/lon in all three measurement systems. The equator becomes an obvious point for measuring latitude. It is defined as 0 degrees while points North are measured as degrees on North latitude until you reach 90 degrees at the north pole. Similarly the Southern Hemisphere is measured in degrees of South latitude. There is no corresponding obvious point for longitude lines. By international agreement 0 degrees longitude is a meridian line that cuts through England and is called the prime meridian. Longitudinal distances are measures west and east until they come together at the International Date Line 180 degrees later.

While Garmin can display and convert from the three different systems used to define lat/lon there can be considerable confusion among users looking at the display and comparing it to some external information. This is because some data is not very precise in its use of decimal points and the lack of a degree sign on most keyboards can encourage a substitution of the decimal point. When comparing numbers, consider that minutes can only go to 59 and will then roll over. If the data just after the decimal has digits above 5 then it is likely to be a decimal part of a degree and not minutes. Similarly, this line of reasoning can also be applied to seconds. A space between degrees and minutes is the preferred separator to avoid confusion when a degree sign cannot be used. The ' is used to mark minutes while the " should be used to mark seconds.

Decimal parts of a degree can be converted to minutes easily by multiplying by 60. Similarly decimal parts of a minute can be converted to seconds by multiplying by 60 thus, if you have ddd.ddddd you can convert to ddd mm.mmm by:

$$ddd \ mm.mmm = ddd + (0.ddddd \times 60)$$

or conversely

$$ddd.ddddd = ddd + (mm.mmm / 60)$$

Many folks think the lat/lon measuring system is independent of the datum issue however this is not the case. In the precision and accuracy required for GPS use the same problems are found with lat/lon as with any other grid system that might be used. It earlier times measurements weren't precise enough to make this obvious. Even today making angular measurements using devices like sextants is not accurate enough to make the datum usage much of an issue.

Angular measurements, however, are not particularly convenient for measuring distances and, even in its simplest form requires the use of trigonometry. Long distances become even worse since a line between two cities at different longitudes and latitudes on a map does not represent the distance traveled between those two points when actually traversing the distance on the real earth. Instead we have to resort to great circle distances and measurements for these cases. Mariners and aviators use nautical miles to help simplify this calculation. One nautical mile is 1 minute of latitude thus a degree of latitude is 60 nautical miles. When measuring distance in the east-west direction the distance correction varies with latitude. While accurate measures would require us to consider the earth as an ellipse a simple estimation can assume the earth to be a spheroid and thus the east-west distance in nautical miles is given by the formula:

distance = (difference in minutes) x cos(latitude)

Thus a degree difference in longitude at the equator is 60 nautical miles but at 45 degrees latitude it would only be 42.426 nautical miles. For this reason other grid systems have developed to permit more direct measurements to be made on maps. One recent world wide grid system that attempts to solve this problem is called UTM.

UTM

Let's suppose we took an orange and cut the peeling vertically around the orange in several places and then peeled it off and laid the pieces side by side. Then we took a hammer and flattened the whole thing out. We might have something that looked a little like the image below:

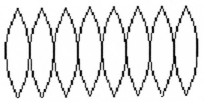

Figure 12 Mercator image

This image is similar to the projection system we get to create the UTM grid system. The UTM, Universal Transverse Mercator, system does a similar thing with the earth by slicing vertically every 6 degrees thus creating 60 such slices around the world. All the slices touch at the equator and get further apart as we get closer to the poles. Another interesting thing to notice is that the longitude lines at the center of each slice are straight while at the edges of each slice they are curved. The amount of curvature increases as we get closer to the edge. A vertical line drawn near the edge would not point directly north.

Each slice, called a zone in UTM, is given a number from 00 to 59 starting at the International Date Line and progressing east. While not necessary to the measuring system UTM also divides each zone horizontally as well. These divisions start at the equator and are 8 degrees wide. The first half of the alphabet is used for the southern hemisphere while the second half is used for the northern. Thus a point just above the equator would be the letter N proceeding to the letter X. While, just below it would be an M and proceeding backwards to the letter C. In addition to the letter assignment, UTM also measures the distance from the equator in meters along the central meridian. (This is the only straight meridian line in the zone.) Since every zone is measured directly from the equator the letter is not necessary to define the location except in some special cases described below. East-west measurements within a zone are always referenced to the central meridian.

The full UTM grid system is now defined. There is one latitude grid line, the equator, and 60 longitude lines, one in the center of each of the zones. Further each zone is 3 degrees wide on each side of its central meridian. At this point we quit worrying about the angular measurement system and instead just measure direct distances in meters from these lines. Since measuring from a line implies positive and negative values the designers have tried to simplify this problem by defining the central meridian to arbitrarily always be 500,000 meters. This is called a false easting. This means that distances to the left of the central meridian will be subtracted from 500,000 and those to the right will be added to 500,000. In this way all measurements will always be a positive number of 6 digits. North/South distances are measured from the equator directly and can get as large as 7 digits. For this reason we usually add a leading 0 to the east/west distances to make them 7 digits long also. To avoid negative distances in the North/South direction in the Southern Hemisphere always add 10,000,000 meters to this measured negative distance.

Since each central meridian is always 500,000 meters there needs to be a way of designating which central meridian is being talked about. This is accomplished with the zone prefix (a number between 00 and 59) thus a full UTM specification consists of a two digit zone number, the zone letter and 14 digits of measurement data to measure down to the level of one meter anywhere on earth. The first half of the measurement data is the east/west number while the last half is the north/south component. If you don't need the full precision of 1 meter you can leave off pairs of digits. You could use 12 digits for 10 meter, 10 digits for 100 meter accuracy, etc. Sometimes you will see an odd number of digits by leaving off the leading 0 in the east/west number.

Since there are so many numbers there needs to be some method of separating them out. In standard usage a comma or a point is used to separate the numbers into groupings. Map makers have chosen to vary the font size. For example instead of 0392000E and 3382000N you would see something like 0392000E and 3382000N. This is particularly important alignment information if you leave off some of the digits.

UPS and special zones

As can be imagined from the drawing above the tips of the points have very little area associated with them and a great big gap. UTM solved this problem by drawing a line at 84 degrees North Latitude and 80 degrees south latitude. The areas above and below those lines are rearranged into two circles like pieces of a pie. These new zones use a different pie shaped grid system called UPS. The southern pie is a bit larger since it needs to cover Antarctica. Two of the letters on each end of the alphabet are used to reference zones in this area. A and B divide the southern pie into two hemispheres while Y and Z do the same with the northern. The map picture for this area would be a polar projection.

A few of areas below the 84 degree separation also have been modified. This has been done to permit an island to be fully contained within one zone. At these latitudes you won't have any problem overflowing the 50,000 false easting even with the wider zones. Zone 32 has been widened to 9° (at the expense of zone 31) between latitudes 56° and 64° to accommodate southwest Norway. Similarly, between 72° and 84°, zones 33 and 35 have been widened to 12° to accommodate Svalbard. To compensate for these 12° wide zones, zones 31 and 37 are widened to 9° and zones 32, 34, and 36 are eliminated.

GPS helps UTM

One of the biggest problems in using UTM is the handling of the zone splits. Within a zone it is very easy to find coordinates and to measure and compute any distance on the map. When you hit the boundary you are suddenly confronted with a different number with no relationship at all to the one you just left in the east/west direction. When a map has a populated area that crosses a zone boundary this is usually handled by placing a secondary grid in the new area using the old zone grid data. This secondary grid permits measuring and locating objects on a map based on a projection from a location in the other zone. Thus in areas near a zone boundary you will see both zones extended into the other zone area. These other zones will not be aligned with the edge of the map that makes them easy to spot. Zone boundary lines are usually drawn right on the map to help you with this.

Your Garmin unit automatically knows when you leave one zone and enter another. Further if you project a waypoint into another zone it will be automatically recalculated into the correct value for that zone changing the zone number and letter as needed. In addition you needn't enter the zone letter for any waypoint you need to enter except that, because of the false northing for Southern Hemisphere you will need to be sure the letter code is in the correct half of the alphabet. In addition you will need to use an UPS zone letter to let the GPS know when you want a grid in this zone.

UTM is the most precise measurement system on your GPS. It reports your position to 1-meter precision. None of the other grid systems can achieve this level of precision.

MGRS

Some Garmin GPS receivers also support the Military Grid Reference System. This system is just another form of UTM so if your unit doesn't support this it is easy to translate to MGRS from the UTM numbers. MGRS replaces the most significant digits of the UTM coordinate with two letters. If you use the example shown above for UTM 0392000E and 3382000W you would see UQ 9200082000. The two letters UQ are in addition to the normal zone number and letter code. The two letter codes of MGRS are not unique and may repeat at other points in the globe but they are unique enough that for tactical use within a few thousand miles the zone number and letter code are not needed. The two letter codes are clearly indicated on the map and are used as the main reference locator. These two letter local square designations permit rapid orientation and rough position indications.

In the same way that was described for UTM you can use less digits for less accuracy when desired. For rapid orientation to a given area something like UQ 920820 could be useful. For GPS use you will need to enter the full digit code down to the 1-meter level.

Maidenhead

Maidenhead is an angular grid reference system used by ham radio operators to provide a rough indication of their location. It is a supported grid on some Garmin receivers.

The Maidenhead system is a "read right and up" system starting at the 180 longitude and South Pole. Its format is two letters, two decimal digits, two letters, and can be extended for a more exact description of a location. The first letter is 20-degree increments in longitude (A-R), the second 10 degree increments in latitude, and these define a FIELD. The first digit is 2-degree increments in longitude, and the second digit is 1-degree increments in latitude. American hams stop here and call it a SQUARE. The next pair of letters are 2/24 degrees in longitude and 1/24 degree in latitude, and so on. European hams generally use six characters. Garmin receivers display the letters MH for Maidenhead and the 6 characters defined above.

Local Grids

All of the other grids that are defined in your GPS are local grids. This means they are only defined for a given part of the world and cannot be used outside of that context. Attempting to set values outside the defined area will result in blank (underscore) values in the display. One exception is the Loran grid that can be redefined for each area of Loran coverage but is otherwise similar to a local grid. You use a local grid when the maps you intend to use are using a local grid. Otherwise they have no particular use. You can only create waypoints and talk about locations in the context of the area covered by the grid. Most of the time your GPS knows the limit that the grid is defined for and won't even display coordinates for an area outside the grid boundaries. In addition local grids generally require a local datum so you will need to set both.

The question arises as to why there are local grids anyway since the UTM global grid covers the whole world. Usually local grids were defined before the invention of UTM and are thus legacy grids. In addition a local grid usually supports a contiguous set of coordinates in the area defined by the grid. For UTM the coordinates jump every 6 degrees around the world and also at the equator. In addition a local grid was likely established with a local datum and will fit a given local area better than a global grid. For these reasons a UTM grid may not be as useful as a local grid depending on where you are located and the paper maps you are referencing may only support a local grid.

Loran Grids

Loran (LOngRAngeNavigation) is a system that provides location information for mariners. It has been around for many years and is based on some techniques that are similar to a land based version of GPS. Basically there is a master station and 2 to 4 slave stations. Readings are taken on 3 stations and the time delays from the 3 stations are using to triangulate a fix. The stations are usually several hundred miles apart and there are several sets (28) of these stations to cover the entire USA. The master station is one of a chain of 28 stations and is identified with a chain number. It may also serve a slave station for another chain. The slave stations are designated V, X, Y, Z or Victor, Xray, Yankee, Zulu. A Loran grid is special in that it is calibrated to a specific Loran master station and a couple of secondary stations. A different grid could be developed using the same master and a different group of secondaries.

When using this grid you will need to specify the Loran Chain Number and two secondary stations by name, as defined above. You can usually get this information directly from the paper map you are trying to match. If you are in the range of a Loran chain you can set up your grid to a Loran grid and the Garmin will generally choose the correct setup for you. However there may be some overlap so your map might be dependent on a different chain or secondary station so you need to check. The chain number and secondary stations are listed on the left side of the Loran display. If you leave an area defined by that set of stations then the location information will show all zeros. Realize, of course, that this is only a translation of GPS location and the Garmin actually does not use the Loran data.

User-Defined Grid

Many Garmin units support a user-defined grid. This can be used to support grid systems that are not supported directly by the receiver. Unlike the user defined datum however a user-defined grid may or may not be successful. For example the map that contains the grid you are trying to emulate may be using a different projection than the one assumed by the user-defined grid. In addition the allowed references that are displayed may not match the grid you are trying to emulate. For example you cannot match the typical map grid that has letters running one way on the map and numbers the other since the display always works with numbers in both directions. The user-defined grid is really just a modification of the UTM grid and assumes a Mercator projection. And, like the UTM grid convention it does not work well with negative numbers so any grid defined should ensure that the numbers will always be positive.

To define a grid you will need to specify the grid origin as lat/lon and then you can specify a scale factor for the grid and a false Easting and a false Northing in meters. When you first enter the user defined grid settings you will find the values for UTM already entered into the unit. Change these values to match the requirements of the grid you need. If the projection system is different you can still get close approximation for a given local area. You may need to adjust the numbers as you move further away from the origin where you defined them. Don't forget that you may need a different datum, perhaps a user defined one, for your user defined grid. Often though you will be able to use the work someone else has done to define the grid you are interested in and you can just enter the data without having to design it yourself. There are web resources that can be used to define some of the grids you may want.

An example

Suppose you want a local grid of your own that measured distances in feet, here is how to proceed.

1. Pick an origin for the grid. Note that this must be in the lower left corner (SW) of the area the grid is to cover. You can't have any negative numbers in your grid.
2. Go to the origin point and note the lat/lon numbers.
3. Enter the longitude for the origin.
4. Enter a scale of 3.280839895 (This value converts metric to the international foot at sea level. For metric you would enter 1.0)
5. Enter a false easting of 0
6. Enter a false northing of 0
7. Save this configuration.
8. Go look at your coordinates and write down the numbers.
9. Now return to the user-defined grid and enter the numbers you just wrote down as negative numbers in the false easting and false northing.
10. Save the new configuration.
11. Check your waypoint location which should now read 0,0. Remember negative numbers will not work so if it drifted a bit you may have to move.
12. Use your new system.

Sometimes you can't get to the lower left corner as a starting point. Or, perhaps, your origin just needs to be a location that is not at the corner of your grid reference (Perhaps you live on an island where the origin might need to be out in the water.) so you might want a modification to this procedure. The idea is to use the same trick as UTM does by adding a real false easting and false northing to the above procedure. Instead of just adding the negative numbers in the false easting and northing in the above procedure to make the answer some out to zero you could add an offset to these numbers first such as adding 10000 to each negative number. Then save your new configuration and check your waypoint location which should read 10000, 10000 or whatever offset you chose. This offset will need to be factored into any grid values you use for your work so a simple number is best. Using this offset will prevent the negative values problem.

Other Units of Measure

In addition to the units of degrees and meters that are used to define a map datum and grid system there are many other units that are used and defined for your GPS system.

Time

Time is used in your GPS to calculate position. The time used by your GPS is a special GPS time that is transmitted as part of the satellite message. In addition the satellite transmits information in the form of leap seconds adjustments to permit your unit to adjust the clock display to agree with standard UTC time. (Currently this difference is 13 seconds.) All Garmin units also support the ability to change the time zone so that you can display local time. The etrex, emap, 76, and V also support the ability to have automatic daylight savings time adjustment while in the other units you must change the time zone to accomplish this. These settings are on the main menu, system settings.

Many units also store the time and date of waypoint creation in a changeable comment field and all units store time inside the tracklog. The etrex and emap do not store the times of waypoint creation since altitude information is stored in this area. The time stored is not based on your local time. It is either GPS time on some older units or UTC time on most newer models with the latest software. The difference is usually not significant to most users.

Note that the GPS receiver keeps very accurate time internally within nanoseconds to calculate your position but the display of time is not a high priority task for the unit. Thus the time display could be up to a second later (or even more) than the actual time. It is still accurate enough for most people to use to set their clocks. On the emap you will need to go to the main menu and choose setup and then time to see the seconds display.

Linear measure

Garmin GPS units support 3 different user selectable units for horizontal linear measure. These are kilometers, statute miles, and nautical miles. Vertical units are set automatically from the user's horizontal setting to meters or feet. When using miles some units will revert to feet when distances are small. The displayed units may be different than the internal calculation units or the units used on the computer interface. For example NMEA generally outputs distance in nautical miles, after all the M in NMEA stands for Marine. Nautical units are particularly convenient to use when navigating since 1 minute of latitude change is approximately equal to 1 nautical mile. Some Garmin units offer a fourth unit called yards. This, of course, is also the same measurement system as statute miles but the displayed distances are shown in yards when the distances are less than 1000 yards. This is particularly useful for golfers.

Speed calculations are automatically tied to the selected linear measure for display purposes. Traditional units are used such as miles per hour for horizontal speed and feet per second for vertical speed. Similarly some units measure area and these values are also tied to traditional units based on either metric or statute miles.

Angular measure

Angular measure can also be specified as a navigation preference on your Garmin receiver. Angular measure is used to specify your current direction and bearings to objects. Most folks use degrees but some Garmin units also let you specify mils. A mil measurement divides a circle into 6400 units.

The mil unit is primarily used by the military. It is primarily used to help compute a new direction based on an error in an old direction and is used when aiming artillery. The formula is:

mils = lateral distance * (1000 / distance to target)

For example: a 12-mil change would result in a 12-meter movement per 1000 meters distance. The fine division of a circle into 6400 units may have other uses as well.

In addition to specifying the units you can specify the reference for your angular measurement. All units will let you specify true north or magnetic north. Note that a GPS calculates this number based on velocity information or between two locations. Most GPS units do not have a compass and cannot show a compass heading when stopped but will hold the last setting received.

Some units also support grid north and even a user defined north. Generally you should use true north, but if you are also working with a compass you may need to set magnetic north so that the setting will agree with the compass settings. Some compasses will compensate for magnetic declination and will support true headings so this will not be necessary. If you are using UTM maps and wish to follow bearing taken from the map you may need to use grid north since, as pointed out above, no longitude line except the very center one actually is a straight line on one of these maps. If you don't have a grid north setting there may be up to 3-degree error. You GPS will automatically adjust grid north and magnetic north based on your current location. This is done using built in tables and projection algorithms. You can also define your own offset for North if you wish on some units. This, however, is a static number in this case and will not change as your location changes.

Chapter 6
Working with Waypoints

Waypoints are used to store and remember locations that are of interest to the user. They are often used to store intermediate turns and intersections that help define a route to a particular destination. Similar to the waystations used by pony express riders as stopover points waypoints mark significant places on your journey. In some documents and GPS receivers these may also be called landmarks.

Garmin receivers have differing capabilities in the number of waypoints that they can save. The earliest units could save 250 waypoints while later models can store up to 500. The G-12CX can save 1000 waypoints. In addition some units have additional waypoints that are stored in an internal database. These database waypoints are discussed in the chapter on databases.

Generally waypoint names can be anything you wish. On most unit they must be no more than 6 characters long and contain no spaces but can contain any combination of letters and numbers and some punctuation. The newest units permit 10 character names and these can include spaces. The emap, 76 family, and etrex store waypoints in 3 dimensions by including altitude. All other units only store horizontal position. The 76 family will also store depth data for water if an optional depth sensor is attached.

When you first create a new waypoint the unit will automatically assign it a name. These names are three digit numbers starting with 001 and incrementing each time you create a new waypoint. Even if you rename a waypoint the unit will generally assign the next number in the sequence anyway and will not reuse the numbers until it reaches 999. On some units it will automatically use the lowest available number but and on many others you can force the number sequence to continue from wherever you wish. To force a number sequence just name a waypoint from the number sequence. For example naming a waypoint to 001 explicitly will cause the next automatic number to be 002 unless this is in use. If 002 is in use it will automatically increment until it finds an empty number. Some folks assign a waypoint to a high number to force a block of numbers to be sequential for a particular use.

Capturing your position as a Waypoint

Storing your present location is the usual way to record a waypoint. You simply hit the MARK key, on the G-III, etrex, 76, and emap press and hold the ENTER key, and the waypoint entry form will appear. The location where you were at the moment you pressed the button is already saved in the waypoint information on the form so even if you are traveling at 65 miles per hour down the freeway you can reliably capture a single point with the press of the button. A unique default name is assigned automatically but you can modify this name at your leisure without effecting the position data. Finally you should hit "save" to store the waypoint in your database. The waypoint screen from the G-12 family looks like:

```
Mark Position
-------------
Waypoint
  001    .           - Default name and icon - editable.
N 39 15.395'
W121 02.166'         - Current position.
-------------
Add to route
number __            - Add a new waypoint to a route
-------------
 +/- __._FT          - Used with averaging to show FOM
-------------
AVERAGE?             - Averaging available on some units
SAVE?
```

On some units there is an averaging feature. If you are standing still you can click 'average' on the waypoint save screen and the Garmin will start taking readings to average with the one that was captured when you hit the MARK key. Averaging will continue until you finally select save and enter. While averaging is going on the unit will display an FOM, figure of merit, number to indicate the probable accuracy of the average to this point. On the G-III family and emap you first save the waypoint and then you can average by selecting this option from the local menu on the waypoint display screen. (Please see more information on averaging below.)

If you are using dgps you can still average the position but the improvement will be much less since dgps is already more accurate. Averaging still can reduce some of the effect of a temporarily poor DOP. While averaging is going on you can move to the name field and enter the name you wish so that when you have achieved a stable FOM you want you can save the results.

Man Overboard

All units except the aviation models, the etrex, and emap have a man over board (MOB) capability. This is accomplished by pressing the GOTO key (Nav key on the 76) twice in a row and then hitting enter. This will store a special waypoint at your current location named MOB and will automatically start the navigation features of the unit to aid you in navigating back to that point. A position marked in this way is designed as a safety feature for boat use when someone might actually fall overboard. You may find other uses as well such as when your hat blows off your head while driving down the road. If you want to save that location then you should rename it since it will be overridden by the next MOB sequence.

Accuracy Considerations

The accuracy of a particular fix is dependent on a number of factors. For example if the satellite geometry is poor the solution will be inaccurate. Satellites close to the horizon will have good satellite geometry when coupled to one high unit but they suffer from atmospheric effects that can make the accuracy worse. These factors are taken into consideration when the unit reports the estimated position error, EPE, on the Satellite status screen. You should consult this information to determine just how good the fix is. In addition the G-III family reports a Horizontal Dilution of Precision, HDOP, number near the vertical bars on the satellite status screen. (Some other handheld units can also display this information using undocumented commands; see the chapter on "Secret Commands" for more details.) The HDOP number is unitless where higher numbers indicate a worse fix. For a four satellite solution an HDOP of 1.0 would be considered to be an excellent geometry and anything below 2.0 would be great. When there are more than 4 satellites available that can be used to compute a solution the older multiplex units will pick the best 4 for the computation. They will also track 4 more so that as satellites move or the unit moves into new positions a selection of the best four can be made and switched as necessary. Switching to another set of satellites can, of course, effect the EPE accuracy. The newer 12 channel units automatically use all of the available satellites so they tend to be affected less by this kind of change. Using more satellites is called an overdetermined solution and can result in a DOP of less than 1.0.

The meaning of Garmin's EPE number is not documented and it seems to vary for one model to another and even from release to release. With the latest units it seem to reflect a probability from about 50% to perhaps as high as 63%. That is 50% of the time you will be more accurate than the EPE shown but 50% of the time you could be worse. This, of course, is only an estimate but a more conservative approach would be to double the EPE number and expect to be within that accuracy range about 95% of the time.

All of the data above is reflective of horizontal accuracy. Vertical accuracy is generally up to 50% worse than horizontal accuracy but is not reported in the GPS. This is primarily due to the different DOP for a vertical geometry. Many folks are disappointed with this level of accuracy since they often know their approximate altitude and can judge this inaccuracy much easier than they can the horizontal accuracy. If you only have a 2D fix then the most recent altitude setting is used to compute the horizontal fix and if this altitude setting is wrong the horizontal position can be much further off. If you only have a 2D fix you should check and adjust the altitude manually as required to ensure an accurate fix. This can be set on the position page either directly by displaying and selecting it or on the G-III family and emap from the local menu.

If you augment your GPS with a DGPS beacon receiver you can negate most of the atmospheric effects. As a result the accuracy of a DGPS is improved to 5 meters or less depending on the distance between you and the beacon transmitter. A beacon transmitter uses its precisely known location to correct for receive errors and then sends this correction data to your unit. For DGPS to work you need to be able to receive some of the same satellites as the beacon transmitter so that a differential fix can be achieved. In the US the coast guard operates beacon transmitters which are available over much of the country with an effort currently in progress to cover the entire USA. Many countries around the world also have beacon transmitters. However, if you are in a area not covered by these free beacon transmitters or would prefer a receiver that is not a bulky as the standard beacon receiver/antenna combination there are also DGPS services being offered using FM radio frequencies and even directly from communications satellites. Unfortunately these kinds of services are not free.

Note that DGPS not only improves the horizontal solution but also the vertical solution subject to the same 50% degradation due to satellite geometry.

The latest technology for DGPS is called WAAS (EGNOS in Europe) and is available on the newest units. Please see the DGPS chapter for more details.

Averaging Techniques

This section is a little more advanced than others in this chapter. Skip this section if you're not ready to dig into accuracy issues.

If you don't have a DGPS solution available you may still be able to achieve a better than 17 meter solution (the Garmin spec) by the use of averaging. The idea of averaging is to leave the unit stationary and collect multiple solutions and then average them to obtain a better answer. This can be done with the unit itself, collecting data for later computer or manual analysis. Another choice would be to average by hooking it to a computer and collect the data in real time. If you want to average altitude data you must use an external computer since there are no altitude averaging functions inside most Garmin units. (An exception is the emap and 76 family.)

If you don't have an averaging function or can't stand still long enough to use it and you still want to perform position averaging then use the tracklog. Set the sample interval so that the tracklog won't overflow in the time you want to average and let the machine collect the data. Depending on the model unit you have you will have 768, 1024, or 2000 samples when the log is full. Any of these are plenty for our purposes. The tracklog can be downloaded to a computer for analysis or you can perform an unscientific analysis right on the screen. Switch to the map screen and zoom in as far as you can while maintaining all of the points on the display. Now use one of the techniques described below to place a new waypoint in the center of all of the tracks by visually weighing the distribution.

This trick can also be used on tracklogs collected while moving. Suppose you keep a log of several days trips to work and back. You will be able to visually see a distribution at every turn in the route. Assuming the log doesn't overflow and erase your previous trip you should be able to place a waypoint on the map page that is a reasonable average of the trips. You can selectively turn the log on and off to make this a usable method of obtaining an average using the machine itself. Similarly you could collect multiple waypoints over a long period and visually average them to obtain a better solution or use a computer to analyze the waypoints after downloading them. If you use this technique be sure you take them several minutes apart and don't expect great accuracy unless you are willing to collect a lot of points.

Averaging can be done even with a pencil and paper. Generally setting the grid to UTM can make this a bit easier but the idea is to record the location data every 30 seconds or perhaps every minute for a short while and then average the data you recorded. Even 15 minutes of data will improve the location somewhat. Data recorded over a longer interval or with more separation is usually better so you could record several waypoints over several days or even weeks and average them later.

By the way, if you are watching the FOM (figure of merit) while averaging and it starts to creep up instead of down, you may have just experienced one of these anomalies of a fix that is based on information outside the 95% window. Perhaps, if you have time, it would be wise to cancel the waypoint and start over.

Planning Ahead Waypoint Entry

Waypoint entry can be entered several different ways without actually visiting the location or after the fact from collected data.

Entering Known lat/lon Values

If you know the lat/lon for a position you can enter it directly as a new waypoint. There are several ways to enter a new waypoint but the most straightforward is to follow these steps:

1. Go to the main menu and select waypoint.
2. Select New from the 'on page' menu. (Remember that you can use the up arrow key to reach the bottom of the screen.)
3. Enter the new name and the lat/lon entry. You should have the same datum and coordinate system selected that matches the information you are trying to enter.
4. If your machine supports icons you can select the one that most represents the waypoint you are creating. See below for information on icons.
5. Select done when you are finished.

This diagram is for a G-12 family or older multiplex unit. The G-III family is similar.

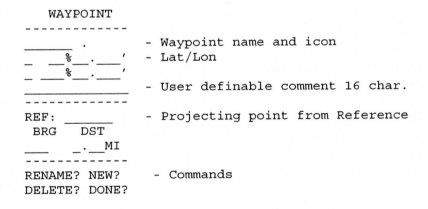

```
WAYPOINT
- - - - - - - - - - - - -
_____   .              - Waypoint name and icon
_  __%__ . ___'          - Lat/Lon
_  __%__ . ___'
_____    - User definable comment 16 char.
- - - - - - - - - - - - -
REF: _____             - Projecting point from Reference
  BRG    DST
____   _.__MI
- - - - - - - - - - - - -
RENAME? NEW?             - Commands
DELETE? DONE?
```

Some units such as the etrex or emap do not have a direct method of entering a new waypoint. On these units you begin by adding a waypoint at your current location, but instead of saving it as you normally would you change the coordinates and then save it.

Projecting a Waypoint

Even if you don't know the specific coordinates you can still enter a waypoint by using the waypoint entry screen to project the new waypoint from an existing saved waypoint on units supporting this feature.

1. Go to the main menu and select Waypoint.
2. Select NEW and pick a name for your new waypoint.
3. Move to the REF keyword, press enter, and key in the old waypoint name you wish to use as a reference for the new waypoint.
4. Enter the distance and bearing to the new waypoint.
5. Pick an icon for your new waypoint.
6. Select Done to complete.

The etrex does support this feature however the method of implementation is different. On the etrex select an existing waypoint instead of a new one. Then select the button "Project" and a new page will appear that looks like a waypoint entry page except that two new entries are included near the bottom of the page. Enter the new name, icon, and distance and bearing to this new waypoint as referenced to the existing waypoint that was used to get here.

Generally the distance and bearing will be obtained by consulting a map. By repeating this process you can enter several waypoint all based on projections from a known point or from each other. It is also possible to project waypoints from your current position. Your current position waypoint name is _____ (6 underscores). Highlighting the REF keyword area, press enter, move the cursor to the left of the left most character to clear the field. This will select the current position as the reference waypoint; proceed as above.

Using the Map Screen to Enter Waypoints

The map screen can be used to graphically enter a new waypoint. You can visually view a location on a map on units so equipped or view a tracklog to determine the optimum setting for a new waypoint. The current zoom setting will determine the accuracy of this kind of waypoint addition so you should set it to as high a zoom setting as is practical. There are two techniques that may be used for entering a waypoint graphically. Both techniques require that you pan to the point where you want to place a waypoint. Just use the arrow keys to pan with the GIII series and emap or enter the pan mode on the other models. Zoom in to achieve the desired degree of accuracy. (The basic etrex has no pan mode and cannot use this method.)

Method one is to hit the mark key and you will get the standard waypoint entry screen. Define the data and press enter to create a waypoint at that point. Here is one area where the G-III family and emap is not consistent with the rest of the Garmin units. Hitting the ENTER key briefly will create a waypoint at the panned position but holding the ENTER key down for long enough to MARK the position will actually enter a waypoint at your current location instead of the panned location. Be careful here and make sure you got what you wanted. In this way you can directly name the waypoint and assign an icon prior to saving it.

Some units may select a point on the map instead of creating a new waypoint. On these units you can convert this mappoint to a standard waypoint by using the local menu and select the convert to waypoint.

The second method for units with a goto key is to pan as above and hit the GOTO key. This will create a waypoint with the name of "MAP". Hitting ENTER at this point will actually cause you to enter a navigation mode with the MAP waypoint as the target destination. However, the MAP waypoint is otherwise like any other waypoint. It will exist until it is manually deleted or replaced with another MAP entry by repeating this procedure. For this reason, if you want to keep the new waypoint you just created you should rename it.

It may also be possible to project a waypoint using the map screen although accuracy may be a bit less. Use the ability to read the direction and distance from your current location while panning to set a new one.

Updating Waypoint Data

One of the things that happens regularly when entering waypoints from maps and other sources is a need to update them when you finally get there. On some Garmin units this is easily done using the

REF keyword on the waypoint page. (The etrex and emap do not have the REF section and cannot use this method.) It can even be done on the fly while driving past the location. Use the following steps.

1. Go to the main menu and select Waypoint.
2. Toggle in the name of the Waypoint you wish to change.
3. Move to the REF keyword area and press enter; ensure the waypoint name is all underscores (move the arrow key to the left until this appears).
4. All underscores is your current location. If you are moving you can watch dynamically as the distance and bearing change to track your movement relative to the waypoint you have selected. (Note that this is a good way to note the distance to the waypoint.)
5. Select the distance field and clear it to zero. (Use the left arrow key again.)
6. Wait until you get to the exact point you want and press enter.

You have just updated the waypoint location on the fly. This is also an excellent way to move waypoints. This technique can be used to shift trackback waypoints to a more meaningful location. You can actually do this on waypoints that are part of an active route.

On units that support averaging an existing waypoint such as the III series and the emap a second way to update an existing waypoint after you arrive is to simply select the waypoint and average it for a few samples. This will move the waypoint to the current position.

Special techniques on mapping units

If your unit supports viewing the waypoint on the map screen you can select it while viewing using the enter key and you will be placed in a move mode where you can graphically move the waypoint.

1. Bring up any waypoint. (It can be selected in panning mode from the map page or in any other method used to view the waypoint.)
2. Bring up the local menu and select SHOW MAP

3. Press ENTER on any waypoint and you will be placed in move mode
4. Use the cursor to move the waypoint as needed
5. Press ENTER to complete the move.
6. You can move others as well.
7. Press Quit to leave move mode.

This is a very handy technique to align waypoints with a road on the map on otherwise fixing waypoints that you have entered via another program. Be careful when using this technique that you don't accidentally move a waypoint. When you are panning to select a waypoint with the cursor keys the selection will happen prior to actually getting the cursor to the exact center of the waypoint. It you press enter the waypoint will be moved so that the exact center of the waypoint is where your cursor happened to be when it was selected. If you were zoomed out at the time you could move the waypoint by a significant amount.

Renaming waypoints and changing icons

Waypoint names can be updated using the Rename function on the Waypoint page if present. Do not just select the name field and change the name. If you do this you will just make a copy of the existing waypoint under the new name. On units without the rename function the waypoint name can be edited directly. All Garmin units except the etrex and emap have a comment line that can also be edited just by selecting it and toggling in the new data. The default comment is the date and time the waypoint was created using UTC time. Often this is what you want but other pieces of information could be street address of the waypoint or altitude. The emap and etrex have an actual altitude field that can be edited in this fashion.

You can change icons, if this is supported on your unit, by just selecting the waypoint icon on the waypoint page. This will bring up the icon page. Select the desired icon and whether the text you wish to be displayed on the map page. Choices include no text, name, or comment. Choosing comment permits you to show waypoint names as long as 16 characters on the map screen.

The different models from Garmin have differing numbers of icons and differing types of icons. The most basic units include 16 different icons that are usually sufficient for most usage. Two will be used by default. All waypoints made by you will default to the small dot icon and all waypoints created by the computer as part of a trackback will use the Temp icon. On the G-12 family and G-II+ this is a small T surrounded with a filled circle. On the G-III family it is a set of footprints. Do not use this icon yourself so that you can readily identify the temp waypoints.

Some favorite icons for the G-12:
dot - used for turns in the road as part of a route.
house - used for all buildings
gas - used for favorite gas stations.
car - used for parking lots and all transportation sites.
boat - used for lakes
exit - used for freeway exits.
box with flag (school house) - used for marking all towns.
Tent - camp ground
cross - used for intersections. (most used icon)
T - temp - only used for trackbacks.

Others include the fish, anchor, wreck, poison, circle x, and deer. You may use these as well, of course. Some folks mark their fishing spots, keep away from (proximity) waypoints, and use the cross for hospitals and the circle X for intersections or train crossing. You need to develop your own conventions and then stick to them. Units that support maps will tend to have more icons and have standard map symbols for icon use. Some have as many as 75 different icon choices.

You can also use the main waypoint screen to change the values present on the waypoint data itself or the comment by selecting and modifying them directly. For waypoints that have a specific street address, this information can be placed here, or perhaps you might want to enter the altitude. Units that support altitude in the waypoint generally do not support comments.

Viewing Waypoint Data

Waypoint information can easily be viewed from the main menu by using the waypoint command. Once entered the waypoint screen will appear with some waypoint selected or perhaps a totally blank screen. To select the waypoint you want to view you need to toggle in its name by selecting the waypoint name field and then using the arrow keys to select the name you want. Use the up/down arrow directions to pick the letter or number and the right arrow to move to the next letter. The Garmin has a feature that attempts to fill in the waypoint name for you based on the first waypoint name that matches the letters you have toggled in so far. However, it is easy to skip over names without noticing. For example, suppose you have waypoints named ANDY, APPLE, BANANA. You enter A and ANDY shows on the screen. If you were to continue to toggle the A to B you would see BANANA but you might not know that APPLE was even available. To reach APPLE you must key A and then move to the second letter and change it to a P to see APPLE.

You can also view a waypoint from many of the screens that permit you to select a waypoint. For example if you highlight a waypoint on the map screen or on any of the route screens you can press enter and review the contents of the waypoint or work with that waypoint. From the main menu page there is a waypoint list menu item. This list can also be used to select a waypoint for viewing or editing. On the 12CX the menu list is augmented with a tab system that lets you get to a particular waypoint more easily.

On units that have a find key you would view a waypoint by using the find key and then selecting waypoint. You can view the full list or the nearest list which supports up to 15 nearest waypoints on the emap. (They must all be within 82 miles to be seen.) There is also a favorite list on the advanced etrex models that lets you build of list of favorites for rapid viewing at any time.

Your Garmin keeps track of 9 waypoints (15 on the emap) that have the special significance of being the closest waypoints to your current position. These waypoints are displayed on the map screen and are available in a special list from the main menu. This can be a big help if you get lost and need to find the closest place near where you are currently located. Waypoints beyond 100 miles will not be shown even if you have less than 9 available.

Some of the waypoints are visible on the map screen. On mapping units you can use a local menu item on the waypoint screen to switch to the map view so that you can visually identify the waypoint. On the emap and etrex this is a button on the waypoint screen. On all units the nearest nine waypoints are visible on the map screen. If you want to view a waypoint that is not in the nearest nine you could use the simulation mode to move your position so that it will be visible. Another technique is to use the 'goto' command to set up a temporary goto. This will make it visible and draw a line from your current position to the waypoint to help you find it. If you are viewing a waypoint, or have it highlighted in a list then this will be the default target for a goto which make this technique much easier. Finally, if you need to view several waypoints that are not in the closest nine you can create a route of up to 30 waypoints. If this is activated they will all be visible on the map page.

The etrex can display all the waypoint on the map screen, which can cause clutter so you have an option to turn some of them off. The emap will display up to 15 waypoints but when you scroll the screen the 15 waypoints chosen will be the closest to the cursor instead of being based on your location as in the other units.

Deleting Waypoints

The rules that permit deleting waypoints is quite a bit different among the various Garmin units. This is by far the biggest difference in these units. Generally you can delete waypoints from the waypoint page by using the delete option but there are conditions that won't let you delete some waypoints. For example, none of the units will let you delete a waypoint that you are currently using as the destination for navigation.

The older multiplex units won't let you remove any waypoint that is part of a route. To remove one of these waypoints you need to first remove it from the route or routes that it is in. On these units the waypoint list includes one route number if a waypoint is in a route. So, you would find the waypoint in the list and note the route number prior to attempting to delete it. This information is displayed where the icons are displayed on the later units. Removing a route or removing a waypoint from a route has no effect on the waypoint itself.

The 12 channel units have relaxed the rules for removal considerably. On these units you can remove a waypoint and it will automatically be removed from any routes that it is part of. The etrex and emap warn you that the waypoint is part of a route and you can then choose to remove it or cancel. You cannot remove a waypoint if it is currently the destination point for a goto or is defining the current leg of a route you are using. On the G-12 family temp waypoints created as part of a trackback will automatically be removed if they are no longer part of any route and a new trackback is initiated.

There are also commands for removing multiple waypoints. At the bottom of the waypoint list there is a "delete waypoints" command. On older units this will remove all waypoints not part of a route. On newer units this will invoke a submenu permitting removing all waypoints or all waypoints that are using a particular icon. On the later releases of the G-III family you can further distinguish removal by specifying all waypoints using a particular icon that are not part of a route.

The GPS V does not seem to have an easy way to delete all waypoints. Here is a work around to delete all of the waypoints. Use the Find command and select waypoints by name, press the page key, then hit the menu key and select delete all.

Distance to Waypoints

You are often interested in measuring the distance to a waypoint or between waypoints. Here are a few techniques. Some also tell you the bearing to between the waypoints too.

- Many Mapping units have a special function available on the map page that can measure the distance between any two points on the map. It is on the local menu.
- Distance from your current location - hit the goto key. The map page and the navigation pages will indicate the distance to the waypoint.

- Distance from your current location - Pan the map screen until the waypoint appears selected. The distance will be shown on the screen.
- Distance between waypoints - Select one waypoint and bring up its waypoint screen. Select second waypoint as a reference waypoint and the distance and bearing between them will be indicated. Note that if the reference waypoint is _____ then this is the distance from your current location.
- Some units have a DIST and SUN entry on the main menu - select this entry and enter the two waypoints. It will also show the sunrise and sunset at the destination waypoint.
- Distance between waypoints - Place both waypoints in a route and the distance will be shown on the route screen.
- The nearest waypoint function will tell the distance from your current location to nearest waypoints, which will also be sorted based on distance. The nearest waypoint may be an entry on the main menu or it may be coupled into the find waypoint command as a menu item.
- Units with a find button can use this function to get the distance to a waypoint. Use the find button, select waypoint (by name, nearest, favorite, or recent depending on your model), highlight the desired waypoint and its distance (and bearing) will be shown at the bottom of the screen.
- Units with a find button can also use this function to get the distance between waypoints. Use the find button as above to select the first waypoint, view the waypoint and choose map to show its location on the screen. Now use the find button again to locate the second waypoint. Distances and bearings will now be shown relative to the first waypoint you selected.

Proximity Waypoints

Proximity Waypoints are the same as any other waypoint in your waypoint database. However, some Garmin units, the G-12 family and the 76 family permit you to use waypoints in a different way.

These units have an alarm that can be set to indicate when you get within a defined radius from these waypoints. Most commonly you would use this to indicate a warning when you got to close to a sand bar or other obstruction when boating. But, there is no reason you couldn't use them to let you know when you were in CB range of you home or within radio reception range of a favorite radio station or perhaps any other use you can imagine. Proximity alarms are separate and independent from the arrival alarms that can occur when you are using goto's and routes.

To set a proximity alarm you would go to the main menu page and select proximity waypoints. You then add a waypoint from your list of existing waypoints and add the distance from the waypoint. Whenever you are within the radius defined you will be alerted. To disable that alarm set the distance to 0 or remove the waypoint from the list. Alarms are both visual (with an alert box) and audible, except on the G-12. At night if the lamp has timed out it will flash on as an additional visual warning.

Tips and Tricks

Tip 1 - Averaging an existing Waypoint
It is not possible to use averaging when updating a waypoint, on the 12 family or II+, since the averaging selection is only on the new waypoint screen. If you have an existing waypoint that you want to update but it has comments or a long name that you don't want to re-enter you can build a new waypoint with a simple name like "a". Average it as much as you need to. Then bring up the waypoint you want to change and set it to reference the new waypoint "a" and set the distance to '0'. Hitting enter will change the existing waypoint to use the new waypoint's position. Now delete the "a" waypoint.

Tip 2 - Making waypoints from the city database
It is possible to create a waypoint from the city database. If you select a city on the map, or map screen, and then hit the "mark" command you will create a user waypoint that is derived from the city one. The emap does not have a mark key but any object on the map can be converted to a waypoint using an entry on the local menu after

you select the object. Note that the emap, etrex, and 76 units can use a mappoint selected from the map in a route without having to first convert it to a waypoint but on older units it must be a real waypoint to be in a route.

Tip 3 - Returning to a previous waypoint

One way to partially defeat any inaccuracies and return to your exact fishing spot is to store your waypoint, averaged as best you can. Then take two bearings of prominent objects using a hand bearing compass or the built in compass in some Garmin units. Store the bearings as part of the comment or in a separate notebook. (Storing just the two bearings should be enough if the prominent landmarks are really prominent.) Now when you return to the spot you can use the bearings to help locate the exact position.

Tip 4 - Triangulation Technique

Let's suppose you can see some point off in the distance that you would like to navigate to but you don't know its distance. With the use of your GPS and an external hand bearing compass (or the built-in compass on the summit an vista) you can triangulate and find this destination waypoint. The method will be to build a route and then use the display of the route on the map screen to pinpoint the desired location. Here is how to proceed:

1. From your current location mark a waypoint.
2. Tak a bearing with your compass to the desired point and use the waypoint editing page to add a new waypoint based on bearing and distance from your current position waypoint. You will just estimate the distance, so pick a distance that is further away than you expect the point to be. (On the Vista and summit you just do a sight n go command and the "Site n go" waypoint is built automatically, rename it for the next step.)
3. Begin to build a route by starting with the new waypoint you built out of the bearing and the second entry should be the current location waypoint. This is just a temporary route so build it on the route planning page on the G-III family or on route 0 for the other products. See the route chapter for instructions on entering a route. Note that

adding a waypoint entry to route 0 will automatically cause the active route page to appear in the rotation so use this page when you need to add further waypoints.

4. Travel to a second point some distance from your current location where you can get another bearing on the distant point. Repeat steps 1 and 2 for from the new location. The further you travel the more accurate the answer is likely to be.

5. Add the now current location to the next entry in the route and finally add the new projection to the route. You should now have 4 locations in the route.

6. On the G-III family you need to activate the route on the planning page but on other units it is already active since you used 0 to build it.

7. Look on the map page and you should see a route that looks something like a child's drawing of an Indian teepee, a base for a triangle and two lines extending from each end of the base and crossing somewhere on the screen. This crossing point is the desired destination.

8. Use the panning capability on the map screen to move the cursor location to the intersection point found in the previous step and zoom in for the most accuracy.

9. Hit the GOTO key to create a new waypoint called MAP at that point. Your GPS will now be navigating toward this destination. Or just use the enter key if you don't want to navigate to that point.

At this point you can get rid of the active route, remove the temporary waypoints and use the navigation screens to goto the new waypoint or to compute the distance to the new waypoint.

Tip 5 Reading current lat/lon and altitude
The emap and etrex will provide information about your current location but it may take several keystrokes to get at this data. It is

often easier to press and hold the enter key to bring up the waypoint screen. This will show the current location and altitude. If no other keys are entered this waypoint can be canceled (esc on emap, page on etrex) when you are finished looking at it. This trick will work on any unit, of course.

Tip 6 Seeing more waypoints on the screen.

Many Garmin units restrict the number of waypoints that are visible to some number. On some units it is 9 (older units) while some have 15 (emap). On these units you can view more waypoints on the screen by making and activating a route. All the waypoints in a route are always visible.

Tip 7 Getting new waypoint icons
Units that support maps and/or poi data may display icons for this data that are not available within the standard icon list for a waypoint. It is possible to use these icons for your own waypoints. Here's how:
1. Display a mappoint (poi) containing the icon you wish to use.
2. Convert the mappoint to a waypoint.
3. Edit the name and location data to the location you wish for your new waypoint but don't touch the icon.
4. Save your new waypoint.

Tip 8 Map Referencing on non-mapping units
Road maps often support some arbitrary grid reference system that may not be duplicated in the GPS grid, such as A1, C6, etc. While it might be possible to define a user defined grid for the map it may be more trouble than its worth. Another technique is to assign waypoints to the center of these grid areas using these grid names. Then the waypoint will appear on the map screen and in the nearest waypoint list, which can be helpful in locating your position on a paper map. To get the waypoints you need to visit one of the center spots or somewhere close and set the waypoint. The rest can be calculated based on projecting from this one waypoint so long as there is a scale on the map or you can estimate one.

Chapter 7
Working with Tracklogs

This chapter covers the use of tracklogs on Garmin receivers. Tracklogs are a very useful feature of your GPS receiver.

Capabilities of a tracklog

Tracklogs are basically the equivalent of dropping breadcrumbs so that you can retrace your steps. They provide a history of your travels. On a Garmin unit they include your position and time recorded automatically as you travel. The etrex family, 76 family, and emap include altitude as part of the log, but earlier units only record the horizontal position. A tracklog can be automatically turned into a trackback route to lead you back to your starting point. In addition they can be downloaded to a computer and used to "playback" your travel over the top of a mapping program so that you can show someone exactly where you went. Since time is also recorded the computer program could also indicate the speed of your travel and compute the length of the trip as recorded in the tracklog. Tracklogs are capable of recording breaks in the log where you moved between fixes such as driving with the unit off or where the unit may have lost a lock. Tracklogs are displayed on the map screen along with visual indications of waypoint locations and your current position.

While all Garmin handhelds keep a tracklog, the length of the tracklog varies from unit to unit. Most older units, 8 channel multiplexing, have a tracklog of 768 points. 12 channel units in the G-12 family and the G-II+ have 1024 points. The G-12CX has a 2,000 point track log. The G-III family has a tracklog of about 1900 points while the G-III and G-III+ (not the pilot) can also save 10 additional tracklogs which are compressed, 256 point max, versions of the main tracklog. These saved tracklogs do not contain the time data. The etrex and the emap, like the G-III family can store 10 saved logs but also offer a unique feature in that these saved logs can be used directly in a backtrack. The basic etrex stores 1535 points in the main log, the emap, most etrex models, and the 76 models store 2048. The etrex summit and vista, hereafter referred to as just the summit or vista, save 3000 points. The saved logs are 256 points except in the summit, which can produce 500 point logs. The saved logs include altitude if it was collected in the main log. The altitude on the summit and vista is recorded from its built in altimeter and is used on the altitude display screen for vertical profiling. The vertical log points on these units will continue to be collected even if the fix goes away, which is one reason that the log is longer.

Note that Garmin originally stated in their specifications and manual that the basic etrex tracklog capacity was 2000 points and the saved log capacity as 250 points. Independent testing indicates a tracklog capacity of 1535 pints and the saved log capacity is 125 points which has been confirmed with Garmin.

Tracklogs need not be contiguous. If you turn your unit off or lose a lock the tracklog will record this as a discontinuous log. For example on a cruise ship you might turn the unit on a few times each day and you will have a highly discontinuous log of your trip. This can still be turned into a backtrack route, which will connect all of the various pieces of the log together into one route, or it can be saved in the saved log, which will do the same thing only you can have more points. This technique can also be used on a hike where you just turn the unit on from time to time to record checkpoints in the hike. Saved logs and routes are always continuous so if you save an active log with discontinuous points they will be connected together to form a continuous log.

Setting up the tracklog

Before you can see a tracklog you must have it turned on. On some units tracklog settings are available from the main menu. On many units there is either a banner at the top of the map page with a selection available or you can access it from a local menu. If there is a banner at the top then the tracklog settings and the map settings are available under OPT, or CFG. Select this choice and then select Tracklog to bring up the tracklog settings. The etrex, summit, and emap always record a log.

You have up to 3 choices in defining how a tracklog is recorded. Some units may not have the fill choice. The etrex, summit, and emap, for example, only record in wrap mode and are always on. Actually the etrex venture, vista, and legend can turn off tracking but it will be turned on again automatically at the next power cycle.

- OFF - Use this choice to avoid erasing a tracklog that you have on the screen that you are using as a map or saving for a later download and don't want to erase it. On units that support an off mode this will be selected automatically if you upload a tracklog from a computer program.
- WRAP - Select this choice to keep the tracklog displayed all of the time. It will keep the latest information in history and will overwrite older data if the tracklog becomes full. This is the right setting if you are most concerned with your destination location.
- FILL - Selecting this mode causes the tracklog to stop collecting data when the tracklog fills up. The unit will give you an alarm to indicate that the tracklog has filled. Use this setting when the original starting point is important to maintain. If you have a download computer (like a laptop or palmtop) available you can use this alarm to alert you to download the tracklog to save it. If you have a G-III or G-III+ you could use this alarm to compress the current log into one of the ten compressed tracklogs. On any unit you can convert the tracklog to a route to save the most important information. Once you have the tracklog saved you can clear the log and continue.

In addition to these choices you will need to decide whether to place the tracklog in "automatic" recording mode or "time" recording mode. In automatic mode the unit itself decides when to drop a breadcrumb (trackpoint). (The G-III family also supports a "distance" recording mode.) Generally, in automatic mode, it will enter a trackpoint when you have turned more than 25 meters (82 feet) from a straight line projection from you last point or you have significantly changed the speed from the last entry. Using these two criteria allows the Garmin to accurately map your journey, however it becomes difficult to judge exactly how much data can be collected before the tracklog becomes full. Some units will also make a log entry when the unit draws a new screen. With a typical 1000 point log you could overflow the log in 40 miles or in 400 miles depending on the terrain and your driving/hiking/riding habits. On the G-III family you can change the setting for the turn distance. The Street Pilot (and probably the V) uses 50 meters by default and this turns out to be a good setting for driving down the road. This, of course, will increase the length of data that can be collected at the expense of accuracy on turns. The etrex vista, legend, and venture have both time and distance choices as well. Automatic mode has a setting where you can adjust the sensitivity to distance from the projected straight line from less to more often.

Automatic mode is the only mode supported on basic etrex, summit, and emap units. These three units have other methods of using the automatic mode. For example, they can use speed to help determine how often a trackpoint is laid down. When walking a departure of 10 feet from a projected straight line might be enough to trigger a new tracklog point but while driving this would not be enough. The users current zoom scale can also be used as an indicator of desired track resolution. Garmin has not documented how they are doing the tracklogs on the latest software releases of these models but empirical data from users has shown that some of these techniques are being used. Expect more refinement in automatic modes in the future as Garmin tunes the products to attempt to produce the maximum significant data with the minimum number of points.

When you need to guarantee a time before the tracklog fills up you should use time mode. In this mode you set the period for collecting information and can thus determine exactly how long it will be before the tracklog fills. Unfortunately this may not provide accurate information about turns in the route since the sample interval may not record this immediately. Time recording is the best setting to use if you intend to record information at a stationary point for later averaging. You can record the tracklog, download it to a computer, and average the datapoints for increased accuracy.

The III and III+ and etrex units with a "click stick" can also set distance for tracklog recording. This is another way to guarantee the log will not fill up prior to the end of the trip. Note that distance is measured from the previous trackpoint and is not based on your actual travel distance so if you drive around in a circle you won't fill up the log as long as the diameter of the circle is less than the distance you set. When you are hiking the smallest distance setting will result in the most accurate tracklog available on a III and III+.

The final setting on the tracklog menu is an indication of how full the tracklog is currently and the ability to clear the log, erasing the current entries. This should be done prior to taking a trip where you plan to use the tracklog to generate a return route. Clearing the route of extraneous earlier logs will provide the most accurate backtrack since less information will need to be analyzed.

Using a Tracklog

A tracklog of a mountain road can make you drive like a veteran of that area. After you have driven the road once you have a recording of all the twists and turns in the road. On the return trip you can anticipate the degree of the turns as they appear ahead. While turns may not be exact the information can still be valuable and certainly as good as the memory of someone who drives that route everyday. In addition you can watch the map display as an indication of upcoming turns that require decisions. You won't have to remember which way to turn at an intersection. In this way a tracklog becomes a very useful map for the trip. Be sure and keep your eyes on the road. You can't drive by looking only at the GPS screen!

You can use a tracklog to visually generate a route manually. You can view the track and add waypoints at critical points directly on the map screen. Traveling the same route several times will cause tracklogs to overlap. This information can be utilized to visualize the effect of atmospheric errors over a period of time and may be useful in placing waypoints directly on the map page to average out the effects of these errors.

A primary purpose of the tracklog is to be used as the source of the automatic backtrack routing capability. Once the backtrack is generated it can be used to retrace your steps from one of the navigation screens. The tracklog is no longer needed to perform your return trip, but can still be useful as an indication of minor turns.

Tracklogs may be used to prove that you went to a particular location, arrived at a particular time, and traveled at a particular speed. All times are recorded as satellite times, not local times. A tracklog that is downloaded to a computer will contain this time data but not on upload. This way you are guaranteed that the log on the GPS represents a real trip.

Working with Saved Logs

On the G-III, G-III+, etrex family, emap, and 76 family there are 10 saved logs in addition to the main tracklog. On these units saved logs can be used to provide updates to the map data for roads or trails that aren't on the map. They can also be used to extend the range of the main log or to save routes that are too long or complicated for the route capability. These logs are up to 256 points long (500 on the summit) and will automatically be reduced to 256 points from the main log if it is longer than that when you generate the saved log. This is done by removing all the points that were recorded due to speed changes and by only keeping the most significant turns in the log. The saved log also removes all of the time stamps.

You can turn the display of each of these logs on or off independently. The etrex can only display one of the logs at a time. On all but the III family the backtrack function is actually done using one of these save logs.

To save a log in one of the named logs you should first collect the information you wish to save in the main tracklog. Then follow these steps:

1. Bring up the main menu and select tracklog and press enter
2. Highlight active log and press the menu button
3. Select Save active log and press enter.
4. The log will be reduced to 256 points if required and saved with a default name of the current date.
5. View the log information by highlighting the name and pressing enter.
6. Rename the log as needed by highlighting the name and pressing enter.
7. Press enter when you are finished

The etrex, 76, and emap have added a nice feature to saving a tracklog. They provide distinct points that can be used to start the log, which will always be saved to the current end of the log. These starting points are displayed as a menu item when you select save to save the active log. There are a few choices that are arbitrary times but most choices are based on when you may have stopped or started the unit. Since a log can span several days you will be offered a time that is the last time you turned the unit on this day or perhaps other times you turned the unit on. You will also be offered to save the full log or a log beginning on a previous day. Select the choice your prefer and save the log. Tip: This can also be used to find out what time you turned the unit on by attempting to make a saved log and reaching this menu to find out the time and then cancel the save.

Once you have a saved tracklog there are several things you can do with it. To work with a saved log you will need to select tracklog from the main menu and then highlight the saved log you are interested in. You can then:

- Press menu to choose trackback, or delete. You can do a trackback route from any of the saved track logs. You can delete the individual saved log or all of the saved logs.
- Press enter to view the detailed log data. You can rename the log from this screen by highlighting it and pressing enter.
- Press enter and then menu to select the display options. You can individually choose to display or not display a saved log.

Note that choosing backtrack on an etrex, 76, or emap will allow using points generated in the saved tracklog itself to perform a backtrack. There is no need to generate an actual route as required on the other units. You can directly traverse this saved log in either direction. There are up to 50 significant points saved in the tracklog that will be used for the backtrack navigation but the rest will be shown on the screen as a visual aid. (There seems to be more on the summit but the actual number has not been determined.) On the emap you can move your cursor over one of them and it will indicate a mappoint named "TURN" with a number. All of the turns have this label and while they can be seen in this fashion they cannot be selected. If you are navigating using a tracklog backtrack the significant turns will be indicated with a turn alarm (causing a beep on the emap if configured on). Significant turns seem to be turns of greater than about 60 degrees.

In other respects the emap, 76, and etrex backtrack navigation behaves like a route navigation that is obtained on the other units. An exception is that time and distance calculations are always made to the final destination since there are no intermediate waypoints. If NMEA mode is turned on these backtracks can be used to drive an autopilot on a boat and the intermediate TURN locations will be sent sequentially numbered to the autopilot.

Uploading/downloading a Tracklog

There are many third party programs available that will permit downloading a tracklog. This permits saving tracklogs under different names and using them with pc programs. Many of these programs produce an ASCII file that you can easily edit to shorten or modify in any way. You can also reload these tracklogs back into the Garmin unit. When uploading a tracklog on most units the date and time data is set to 0 so it is not possible to upload a false track log. At least it is easy to verify if you do. Once you upload a track most units will automatically turn off track recording to preserve the log you just loaded.

You may want to upload a tracklog to restore a road map that you want to use for a particular trip. Or perhaps you could combine tracklogs to have several local roads or other track data. There are even programs available that will permit you to graphically generate a tracklog by tracing over a map. These can be used to build a lake outline or perhaps a shore outline. In this way you can have a simple map in units without mapping capability or update a map in units that have one.

The Garmin III, III+, etrex, summit, and emap will always download the full log including all of the saved tracklogs. You can usually tell where the split is between the saved logs and the main log since the saved logs do not have a timestamp. The saved logs are also named. There are two download/upload protocols available and if the program uses the latest one it will also get the saved tracklog names. Uploading can a bit more problematic since in the original protocol you could only upload to the main log. It was easy to create a download that is longer than the one you can upload. You would need to trim the download file and load it in pieces. Once in the main tracklog you can use the G-III to transfer it to the appropriate save tracklog. Using the latest protocol you can now directly upload into the saved logs by name which makes track log maintenance much easier. Note that the main log is called "active". You will need to check the program you are using to determine which method it supports.

Chapter 8
Working with Routes

A route is a collection of waypoints that are related in a way that permits you to use them to follow a prescribed course. Each section of the course is called a leg of the route. Generally on the Garmin units you can create a route with 2 to 30 waypoints (1 to 29 legs) and the unit will keep track of up to 20 such routes. The etrex, emap, and 76 have up to 50 waypoints in a route. The basic etrex supports one route, the etrex summit supports 20, and the emap supports 50 as does the 76. There are a number of things that can be done with routes, but the most common is to use a route to guide you to a destination.

Actually many people don't see the need for routes at all. If you need to go to a destination that exists in your list of waypoints you can simply use the "goto" command to guide you to the destination. Under these conditions you have created a route of two points, where you are located now and a target destination. Once you have initiated a goto the Garmin navigation screens will point you toward the destination and the map screen will draw a line from your present location to the target location. Of course, you may not be able to head directly toward the destination since there may be obstacles in the way so you must navigate by working your way toward the destination while moving around whatever obstacles you may encounter. This works surprisingly well even when driving in a car where you use the compass navigation screen to keep the destination objective in mind while traveling down the road. A feature of the "goto" command, on units equipped with this key, is that you can re-initiate it at any time by hitting the "goto" key again and hitting "enter" to repeat the same destination. The GPS will recompute a line from the current location to the destination. The emap always recomputes from the present location while the etrex has an option on the local menu so that it can work either way. To force a re-evaluation on the etrex you will need to switch it to current bearing or re-initiate the goto. The GPS will keep an updated calculation of the airline distance to the destination and will compute your probable arrival time based on your current speed and heading.

The newest units have no goto key so on these units you would use an object oriented approach, that is, find the waypoint and then the waypoint page will provide a goto button. The emap and other newer mapping units also support direct goto's to map objects without the need to create a waypoint. Use the cursor keys to highlight the object and then press enter to get information on the object. A goto button will be on the menu. The emap also uses a slightly different method of computing your probable arrival time. Instead of using the current speed and heading and computing a velocity made good it uses the most recent sustained velocity. Thus it does not change the value rapidly nor does it respond to acceleration or even stopping. The value will be changed when you achieve a new sustained velocity. You can cause the value to be recomputed on this unit by using the stop navigation/resume navigation command sequence.

Many times, however, a closer approximation to the path being followed can be a big help. There are people at the other extreme that say the 50 points in a route is not enough since doesn't exactly match every curve in the road and insist that a full tracklog can be used. But, most folks find a route that tells you where the turns are is a great aid to navigation. A route with intermediate waypoints can alert you to turns in the route to keep you from getting lost. Further, an approximation of the actual course will permit better estimates as to distance and arrival times. These intermediate waypoints may also indicate probable stopping points such as rest areas or refuel points. The GPS will estimate time and distance to these waypoints as well as time and distance to the final destination. This estimate is based on your current speed and an expectation that you will be following the route vectors.

Creating A Route

The easiest way to create a route may be to use the trackback feature available on Garmin handhelds. This feature is designed to use the log that was created when you traveled somewhere to create a route that can be followed to return to the origin. To create a route using the tracklog follow these steps. (Trackback on the newest units such as the etrex and emap works with a saved track log and not a route so this technique is not applicable to these units.)

1. Be sure that the tracklog feature is on. The tracklog feature can be reached either from the map page menu under Track Setup or from the main menu page under track log. On some units you will have a choice to collect data using a wrap mode or a fill mode. This is only important if you are likely to exceed the capacity of the track log on your trip. Use wrap mode if the data closest to your destination is most important and fill mode if the data most closest to the starting point is most important or if you wish to insure that all data is available. For trackback purposes the automatic tracklog does the best job of collecting data but use timed mode if you must insure that the log does not overflow. Be sure and clear the log at the start of the trip.

2. Drive to your destination. If your tracklog is set to fill and you happen to fill it up you will get a warning message. At this point you could generate a backtrack and then clear the log and start it up again if you wish to have a complete record of the trip. (On the G-III you can also save the tracklog in a secondary saved tracklog.) You can also monitor the tracklog even if you don't have the warning and decide to generate a backtrack route from any point you feel is significant. A trackback route like any other route can have, at most, 30 waypoints in them. If the route is really complicated you may feel that 30 waypoints are too few when the tracklog is full. Under these conditions you will need to generate the trackback earlier. Note that you can reinitialize the tracklog once while the trackback route is saved in the active route log. Doing this a second time will clear the earlier route without warning so be sure and save it first.

3. Once you have reached your destination, or prior to beginning the return trip, you simply activate the trackback feature to create the return route. All of the routing features and navigation features of the GPS will be available to help you return to the start of your trip. The command location for enabling a trackback is different for the different units. On older units you can find it on the same menu where you found the original track log. On

newer units it can be found at the bottom of the goto menu
that is reached by pressing the "goto" key.

Invoking the Trackback command causes your GPS to analyze the
tracklog and look for the most significant turns in the log. It will then
reduce the number of turns to a maximum that will not exceed the
route capacity of 30 waypoints. (Actuall testing indicates that it never
goes beyond 29.) Trackback routes work by creating so called
temporary waypoints. These are like the other waypoints and use up
waypoint memory just like other waypoints do. They have a special
name that begins with a T and is followed by a unique number and if
your unit supports icons they will have a special icon assigned. If your
waypoint memory is full then the trackback command will warn you
and use the closest match in existing waypoints to create the
backtrack. On some units initiating a trackback will erase all
temporary waypoints that are not part of a route. However, on most
units temporary waypoints are just as permanent as regular waypoints.
The Garmin III says that if it runs out of waypoints in the database it
will start removing temporary waypoints that are not part of a route.
Erasing the route created by a trackback will not erase the waypoints
themselves.

If you have a G-III family unit with multiple track logs you can
also create a backtrack route using one of the saved tracklogs.
Perform the same steps to select a backtrack and then select the menu
item saved logs and follow the messages to select the saved log of
your choice.

Creating a route manually

The easiest way to create a route manually, on many units, is to
use the "mark" key to mark the route as you are traveling. The menu
that appears when the "mark" key is depressed includes an entry
called "add to route __". If this entry is filled in then the new
waypoint will be automatically added to the end of the numbered
route, if there is room. The route number is remembered so the next
time you can easily repeat the process. Just press mark at the
appropriate point and then press enter to save the waypoint and add it
to the route. The waypoint name will be a unique sequential number.
A tape recorder or one of these new digital recorders could be used to

note the location referenced to the waypoint number. Later when you are not so busy driving, hiking, or sailing, you can return to these waypoints and rename them to something more meaningful.

The advantages of creating a route this way are that you can pick exactly the points where decisions need to be made. Also the waypoints you make will be tagged with the time you made them which can be useful it determining average speed and other trip history data. Note the newest units that store altitude with the waypoint do not have comments or store the time the waypoint was created.

Creating a route from a paper map

If you have a map that includes lat/lon numbers or utm numbers then you can create a route using a similar technique to the one described above. Place the unit in simulation mode, select the grid system that matches your map, select the datum that matches your map, and then go to the position screen. You should find that you can edit the positions on the position screen directly using the simulation mode. Use the cursor keys to highlight the position data and press enter to select it. (If your unit doesn't support this then you will not be able to use this technique.) Once you have entered the data for the location hit the mark key as above and edit the screen to fill in the appropriate data. Repeat for each waypoint in the route.

If you have a scaled map but no grid or would rather not enter all of those coordinate then you can use the following technique to create a route.

1. You must have an actual coordinate for the first point in the route. You can create this first waypoint from a map containing the coordinates or by actually visiting the location. If you only have coordinates for the corner of the map then you can create a temporary waypoint at the corner of the map and then remove the temporary waypoint later. Create this waypoint. (See the chapter on waypoints to create this if you don't remember how.)
2. Select the route menu item from the main menu and create a new route page.

On the G-III family you select a new route page by picking it off the local menu. You could use the active route page for this if you wish since it is always present and is designed to be used for planning purposes. The emap and etrex can only reach the route page from the main menu.

On other Garmin units - To select a new route highlight the route number, press enter, and use the up/down arrow keys to find a number that is not in use. Press enter again and move to the first route point entry.

3. Press enter on the blank routepoint line. You will be allowed to enter the name for the first routepoint, pick the name of the waypoint you entered above. (Or use the local menu if you have a menu key.)
4. Bring up the route menu again and select review waypoint on this waypoint name. This will bring up the waypoint editing screen with the waypoint selected.
5. Select NEW and fill in the new name you want. Fill in the REF: field with the name of the earlier waypoint. Using a ruler and protractor enter the distance and bearing under the reference keyword. You will have to scale the map to real numbers based on your current unit preferences. A calculator or set of dividers can be of use here. Selecting DONE will create this new waypoint as a projection of the information of the earlier point.
6. Repeat these steps for each waypoint you wish to create for the route.

Entering routes from the map screen

Mapping receivers can enter routes directly from the map screen. When working with the route screen you can do a "show map" and enter the waypoint. See below under editing route for a special technique to add waypoints to an existing route from the map screen on the III+. You may wish to just enter the start and end points first to get the rubber band and then use the editing technique to add the intermediate waypoints. This technique will let you get an idea visually of exactly how many points you will need to accurately describe the route. The emap can also use the showmap function to select entries from the map screen but does not have the rubberbanding feature.

Any receiver that supports panning can use a displayed tracklog to build a route manually. Simply pan to the points in the tracklog where you want the route point to appear and press the enter key to make a waypoint. Then the waypoints can be turned into a route. This can be automated to some degree by using the active route page and the view map facility to add the waypoints after the first one is generated. Each time you go back from the view map you will be on the route page. For example the a route can be added to the etrex vista using mappoints as follows:

1. Select a starting waypoint and then use the local menu to add it to a new route.
2. You can bring up the new route you just created the first point for and add a routepoint, this will bring up the find menu.
3. Select a waypoint and then show the map and scroll to any place on the map to highlight a map point. Press the click stick.
4. Ok to add it and repeat the process.

Other techniques

If you already have a set of waypoints entered that you wish to turn into a route then you can simply use the route page and enter them in whatever order you wish. (This most often happens if you accidentally erase a route such as the trackback route and wish to reconstruct it.) The steps are pretty straightforward.

1. Select a new empty route from the route form.
2. Move to the first empty routepoint line and press enter.
3. Toggle in the routepoint name (an existing waypoint) and press enter (or select from the find menu).
4. Repeat until the last routepoint name is entered.

If you don't know the names of all of the waypoints you can use the fact that your GPS will display the nine closest waypoints to help you find them. As soon as you can determine two waypoints use the simulation mode to traverse the route and then you can watch the map screen to find the location of waypoints to add to the route. In simulation mode you can select the speed on either the position page or one of the navigation pages and modify it. The unit will automatically run the route at the speed you select. (The speed on the etrex is fixed in demo mode.) If you only have one waypoint name you cannot activate a route but you can issue a goto to that one waypoint and use simulation mode in the same way. Once you reach the desired point set the speed to zero.

Another way to build a route is to look at the map display and create waypoints directly on the map either using a tracklog, a displayed map (for units having maps), or using a database of cities or buoys. If you scroll the map screen you can use the MARK key to mark waypoints from the map display. If you select a database object on the screen you can use the menu to turn it into a waypoint for use in a route. Otherwise this technique is similar to the manual technique. On mapping receivers creating a route using the map display is even easier. You select New Route from the route menu and then select map display. You can use existing or new routes and they will automatically be added to the route you are building.

The emap, etrex units, and 76 series can build a route using map objects as well. Generally you would start on the route page and then select the object using the find menu for inclusion in the route. Any of the objects can be used and you are not required to convert them into waypoints to use them in a route. If you wish to convert them you can use the showmap command to view them on a map and then a local menu will permit converting them to an actual waypoint. This might be desirable in the emap if the object is on a cartridge that is not always present. If you wish to select the map object graphically then begin on the route page, use find to get to an object that is close to the one you want, and then use showmap to view this object. Once the map is on the screen you can use the arrow keys to select a different object. Once you hit esc to return to the route screen the new object will be the one used. Note that map objects, called mappoints, do not come out of the 500 waypoint database unless you choose to convert them to waypoints. You can have up to 1000 mappoints in use in routes and tracks.

The III Pilot can use database objects from the Jeppesen database as part of a route or as the direct destination. Holding down the goto key on this unit will bring up the Jeppesen database objects similar to the find key on the emap.

The 76 can use points in its city database or navaid database for routes creation.

Some folks never, or seldom, create a route directly on the unit but use a program on their home computer. Once the route is created on the computer an interface cable can be used to download the resultant route onto the GPS itself. There are many programs available that can be used in this fashion and the Garmin MapSource product can be used in this way as well.

Naming the route

No matter how the route is created you will probably want to have a name for the route. On multiplexer units, the II+ and the 12 series the routes are only referenced and used by the route number. However a meaningful route name can keep you from having to remember all of those numbers. Routes that are created using the backtrack feature are automatically named backtrack which is fine for temporary use but not too good for the long haul. You can move the cursor to the comment line and enter any name you wish or you can let the machine automatically name the route by clearing the current name. If you let the machine name the route it will use the first waypoint name followed by the word 'to' (a '-' on the mapping units) and the last waypoint name. If you already have a name assigned and wish to use this feature you will need to erase the current name. Select the name and then move the cursor to the left of the left most character. This will erase the entire field. When you press enter the new route name will be generated automatically. The advantage to letting the unit generate the name is that it will automatically be changed if you reverse the route.

Using A Route

If you only work with backtrack routes then using routes is really easy. You create a route when you need it. The route page will automatically appear in the rotation (G-III family has it in the rotation always) and you use the route until arriving back at the start point. You then clear the route or perhaps just overwrite it with the next backtrack.

Other routes are almost as easy. Select the ROUTE command from the main menu and then select the route number or name. Move to the route you want to activate, select it and then move to the activate (ACT) entry and select (use the local menu on the G-III). Routes can also be reversed using the INV, invert, entry so that the same route can be used to go either direction. The active route page is really just route number 0 so when you activate a route the route you chose is copied to route 0 overwriting whatever used to be there. In all other respects route 0 behaves like any other route. Units that use named routes will allow you to select the route from a list and do not have a notion of a special route 0.

Each time you activate a route it will overwrite whatever route might be present on the active route page. If you want to save an active route you must copy it to another route number before overwriting or clearing it. To copy the active route perform the following steps.

1. Bring up the route command from the main menu page.
2. Select route 0 and then select "COPY TO" by pressing enter on the underscore next to the command. (On the G-III select the TRACKBACK route and use the copy command from the menu.)
3. Use the up/down arrow keys to select an empty route number and
4. Press enter to perform the copy. (Route 0 remains unchanged.)

Whatever method you use to activate a route the GPS unit will use your current location to compute an entry point into the route. It looks at each leg of the route and then projects it into a line and computes a direction that will intersect the closest line. For this reason it will never select the first waypoint in a route as the target for a leg. The navigation screens and map screen will reflect the fact that a route is being used. The map screen will display the route and two new numeric entries will appear identify the bearing and distance to the nearest routepoint on some receivers. Similarly the navigation screens will indicate this data as well as the name of the next waypoint. (See the navigation chapter for more information on using navigation screens with routes.) In addition you will have the route displayed on the active route page. It looks something like the ones shown below.

```
   ACTIVE ROUTE           or    HOME-GAS      ROUTE
-------------------            ----------    PLAN
HOME TO GAS                    WAYPNT <DST>
-------------------            HOME      _._
WAYPNT   ETE   DST             LAKE     2.18
HOME     __:__ _.__            TRAIL    4.05
LAKE     33:04 2.18            GAS      7.24    TOTAL
TRAIL    01:01 4.05            _____   _._     7.24
GAS      01:50 7.24
_____   __:__ _.__            (commands are on a
_____   __:__ _.__            local menu)
_____   __:__ _.__
_____   __:__ _.__
-------------------
CLEAR?  INVERT?
```

If you have a later model GPS the bottom line might read: CLR? INV? ACT? and if you have a unit from the G-III family the table has only one entry instead of two. Of course the names will be different as well. The entry data in the center of the active route table is scrollable to the extent needed to display all of the waypoints in the route. It will be adjusted dynamically to display the current leg at the top and the next few waypoints for as many as there is space to display.

In the example above the field marked ETE is actually customizable. It can be toggled by selecting from three choices. ETE displays the estimated time enroute given the current speed. You can also choose ETA which will display the Estimated time of arrival to the next waypoint (Scroll to the last waypoint in the route to see the ETA for the entire route.) or DTK which is the Desired Track (Bearing) to reach the next waypoint. The DST field shows the cumulative vector distance to the waypoints computed in your current units and assuming you are actually following the route.

The G-III family shows only one entry but the choice of what that entry can display includes much more information. In addition to the 4 entries above, you can check to see if there is sunlight (sunrise, sunset times) at each of the waypoints, fuel usage to that point, fuel usage for each leg of the route, distance between each waypoint, and estimated travel time to that point. This is in keeping with the fact that this is also intended as a planning screen. The actual route screen looks like figure 13.

HOME TO WORK		Active
Waypoint	◄ Distance ►	Route
T004	----ᵐ	
►T005	6.45ᵐ	
T006	6.67ᵐ	
T007	6.99ᵐ	TOTAL
T008	7.39ᵐ	18.1ᵐ

Figure 13 Route Screen

While following an active route the GPS will automatically sense the next leg of the route and switch to guide you on the new leg. This is done mathematically by computing when you cross a line that is projected halfway between the course you are on now and the course for the next leg. You don't actually have to reach the exact waypoint for the switch to occur. Please study the figure shown below:

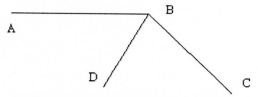

Figure 14 automatic leg switching

Consider that you are following a route to get from point A to point C. The leg AB and leg BC that are part of this route with a common waypoint at point B. The unit projects a line through points BD that bisects this angle. Once the GPS detects that you have crossed this line it will switch from navigating to point B to navigating to point C. Thus the switch will be made even if you are a considerable distance from point B.

The newer etrex models and the 76 family have the ability to turn off automatic switching to the next leg. There is a route setup local menu item on the route page that can be changed to manual. If manual is selected then the user will select the target waypoint for each leg of the route in the following manner:

1. Activate the route
2. Press and hold the find key to get to the active route page.
3. Use the zoom in and zoom out keys to move up and down the list of waypoints to select the target waypoint.

If you have alarms turned on and arrival alarms set to automatic you will receive a warning when you get within one minute (15 seconds on the etrex and variable time based on speed on the emap and V) of each waypoint based on your current speed toward the waypoint. On some units it is possible to have arrival alarms turned on but set to a specific distance. In this case there will be no alarm sounded for the intermediate waypoints but the unit will still switch automatically to each new leg. Routes are turned off automatically for etrex and emap when you reach the final destination but are not turned off automatically on other units and will continue to point to the final waypoint in the route even if you drive past that point. You can clear the route by selecting the clear (CLR) command from the active route page menu. This will remove the route from the active route status but will not remove the waypoints used in the route. On the G-III family there is a deactivate command which will leave the route available on the route planning page. On the emap and etrex you can use the local menu selection "stop navigation" to do this. If you power off while stop navigation is selected then the active route will be stopped permanently. Otherwise you can use the "resume navigation" to restart it on the emap.

If you intend to go back the way you came you might prefer to just invert (INV) the route rather than clearing it. This will reverse all of the waypoints for the route and compute a new entry point into the route. Inverting an active route twice is a way to force the GPS to recalculate its entry point in the route which can be useful when you have left the route for a side trip and wish to return but not to the exact point that you left. The G-III family has a reactivate command on the menu that performs this function so you don't have to invert twice. The emap can use the stop/resume navigation command sequence to do this.

If you can't get the unit to aim at the correct waypoint or you wish to take a side trip you can force the unit to use a particular waypoint using the "goto" key or goto function. Setting a goto will override the current route setting and allow navigation to that waypoint. Once the waypoint is reached the route navigation will automatically continue except on the etrex model where you will have to restart the route.

Some of the Garmin units (the G12 family) also have an activate (ACT) command on the active route page. This can be a very powerful and very dangerous command. The best way to understand how it works is to consider an example. Suppose you selected route 4 and activated it. (This copied the route to route 0 and shows it on the active route page.) It is possible to edit the route on the active route page. If you want to re-copy route 4 and reactivate it while you are navigating the active route you can select this command from the active page. If you had made any changes to the active route they will be lost. So one use of this button is to reactivate and recompute the active route. If you were to switch the route display to route 5 after activating route 4 and some time later decide to select the activate command on the active route page you will automatically switch to route 5. In this way to can easily have a route that consists of up to 60 waypoints or parts thereof. Just pick a route number, activate it, and then pick a different route number. Later when you are finished with the first route selecting this single command will start the second route and enter the route wherever you need to be in the second route.

While this feature is very powerful it can also be dangerous. If you were to have a backtrack route enabled that you haven't saved yet and accidentally selected this command on the active route page you will overwrite your trackback with whatever route was last selected on the route menu. There is no warning. To prevent this you might want to ensure that the last route selected on the route page is route 0 since reactivating route 0 won't erase anything on the active route page. (If route 0 is the current route then this command will behave like the G-III reactivate command.) Taking this precaution is especially recommended if you plan to directly edit the route on the active route page.

Editing A Route

As has already been mentioned, a routepoint and a waypoint are the same (except that the emap and etrex can use a mappoint as a routepoint). If you edit the contents of a waypoint, its location or its name, this information will automatically be reflected in any route that uses this waypoint. If you delete a waypoint it will automatically disappear from any route that used it. Note that early Garmin multiplexing receivers won't let you delete a waypoint if it is used in a route, some other units may warn you but let you do it anyway, but some units will delete it without any such warning. Before you get into trouble you should try this out on a temporary route to see how your particular unit performs. Note that you may want to repeat this test if you do a software upgrade. If you have a multiplex unit you must delete the routepoint first and then you can delete the waypoint. When working with waypoints that were set by the automatic backtrack function you may like to move them slightly to correspond with a freeway exit or some other map identifiable feature. This can be done on your next trip by watching the approaching waypoint on the map screen and noticing that it is close to some map identifiable boundary. Then you can select it and move it (using techniques described in the waypoint chapter) to the exit. The route is automatically updated by this procedure.

Normal route editing commands are reached using the local menu or are available at the bottom of the screen for some commands in some units. Using editing commands you are able to review the contents of a waypoint, delete or insert a new waypoint, exchange one waypoint for another in the route, delete the complete route or copy the route as the starting point for another route. These editing commands are available for regular routes and for the currently active route except that you can't delete a routepoint that is currently your target destination. Deleting a routepoint from your route does not remove it from the waypoint database. To use the routepoint editing commands just use the cursor keys to move to the routepoint you want to work with and bring up the local menu (Either the menu key or the enter key depending on the unit). The choices will be displayed and will effect the current routepoint. If you already have 30 (50 on the emap and etrex) routepoints in your route then you will be unable to select the insert command otherwise selecting this command will shove all of the following waypoints down to make room for the new one.

The "copy to" command will permit you to copy the current route to a new route number (on the G-III family it is the route name with an added number). Select the route you want and then select the copy command. You will only be given choices of empty route locations. This feature is most often used when saving a backtrack route. It can also be used if you want to edit a route by beginning with an existing route. Trackback always creates new routepoints if there is room in the waypoint memory. This is not efficient so you will probably want to change them for existing waypoints if part of your route covers the same ground you have used in another route. To do this easily, activate the route and then run it in the simulator mode, since it is difficult to remember the names for these routepoints. As these double waypoints show up on the map screen, select them and note their names. Then go to the route screen and change the entry for the one you want. (You need to run the simulator or actually traverse the route to get the waypoints to appear on the screen since only the closest 9 will appear.)

Adding route points from the map screen on the G-III

You can add new waypoints to a route directly from the map screen on the G-III family. First, make sure that the ACTIVE ROUTE LINE option is ON for the level of zoom that you want to work in. (The ACTIVE ROUTE LINE option is set in the MAP SETUP SCREEN.)

1. Bring up the main menu. (Hit the MENU key twice.)
2. Select ROUTE on the MAIN MENU.
3. Select an existing route. (It must have less than 30 routepoints.) This will place you in the ROUTE PLAN screen.
4. Select SHOW MAP from the local menu.
5. On the screen you will see the Route Waypoints connected together with the Solid ACTIVE ROUTE LINE. If the ACTIVE ROUTE LINE is not showing, you either have that option turned off or are zoomed out to far.
6. Move the cursor using the rocker key until you have highlighted the ACTIVE ROUTE LINE between two waypoints. The solid line will turn to a dashed line.
7. When the solid line turns to a dashed line, press the ENTER key. The GPS will now display INS by the cursor. This puts you in the automatic insertion mode.

8. Move the cursor to where you want to insert the new waypoint. The dashed line will rubberband along with your cursor.
9. When you have the cursor where you want to place the inserted waypoint press ENTER again. The NEW MAP WAYPOINT screen will then be displayed. This allows you to change the name, symbol, etc. The word USE will be highlighted. Just press enter if you don't want to make any changes. The waypoint will automatically be inserted in your route between the two waypoints connected by the line you picked.
10. When you press ENTER on the USE option, the map will be redisplayed. Simply point to another ACTIVE ROUTE LINE and press ENTER to get the INS by the cursor. Again place the cursor where you want the waypoint and press ENTER again to insert that waypoint.

Unusual Uses For Routes

While generally you build and use a route to help you find a destination there is no reason that this is the only thing you can use a route for. There are a few other uses described in this section. Note that in order for any route to be visible on the map page it must be activated. Normal use of the navigation screens would seem to be disabled when using routes in these new ways but the "goto" route override feature can be used to provide navigation capability.

- One use for a route is to provide a crude map that will be shown on the map screen on units that don't have maps or don't have enough detail. A route might show the location of a highway giving you a visual indication to maintain orientation.
- It is also possible to use a route to indicate the approximate outline of a lake or perhaps a shoreline. This can be handy to force the display of marinas and other features that could be beyond the range of the nearest nine waypoints.
- A route could also be used simply to provide a record of somewhere that you traveled. Later when you returned

home you might use your GPS in simulation mode to retrace the route while having the unit hooked up to a mapping program on your computer. A mapping program will be fooled by the simulation mode into thinking you are actually working live.

- A route could be used to aid in forcing a particular orientation of the map page. Since the display on the Garmin map page is rectangular it tends to provide more information in one direction than the other. Some folks might like to force the long direction to be the favored direction for looking ahead of the current path.
  1. Create a waypoint exactly east and exactly west of your home location
  2. Build a route out of these three points.
  3. Then set the map options to display DTK up and enable the route.
  4. Now the screen display will orient in general agreement with the edges of a paper map that you using and give you a slight edge to seeing upcoming waypoints.

- Similarly North and South waypoints could be set if you prefer this technique to the simple 'North up' setting on the map preferences. A side benefit is that it makes it easy to always know how far you are away from home.

- A route can be used as a planning tool. While in general this statement is true of all Garmins it is particularly true for the Garmin III family. There are special features in this unit to set up average speeds and average fuel consumption and a date and time to permit estimating arrival times and important stop points based on time of day and distance for the planned trip. A few of these features can be emulated in simulation mode on other garmin units by inserting an average speed for the simulation and reviewing the information on the active route screen. All Garmin units except the basic etrex can calculate sunrise and sunset times at a target destination for planning purposes. Note that the time zone is not adjusted for this calculation.

- A route can be used to highlight a destination. For example several folks in the sci.geo.satellite-nav news group used their GPS' to help them find the path of a solar eclipse. Using data provided in a magazine you can create a route of the center line of the eclipse and then just head for a point anywhere on the route. You can visually see the path of the eclipse directly on the map screen. You can even get more exotic by defining a route of the totality drawing two parallel lines with a connection at one end. A route defined this way gives you a visual indication of the location of the total eclipse. And, if you are using a GPS with built in maps, you can perhaps spot a viewing location that is a little more out of the way than the places the crowds know about.

- A route can be used to force the waypoints to appear on the map screen. Most units only display the 9 closest waypoints on the map screen (15 on the emap) and then only if they are within a fixed distance. Activating a route causes all of the route waypoints to appear in addition to the closest ones.

- A route can be used as a waypoint management tool. For example, it can store a collection of waypoints of places you intend to visit on your vacation or some of your favorite fishing spots. In this case you might never actually activate the route but just placing the waypoints in a route provides a way to easily find them as a collection and you can use goto to actually navigate to one of the locations.

- See the waypoints chapter tips 4 and 6 for a couple more uses.

Chapter 9
Working with Autorouting Receivers

Autorouting is a feature where you can simply request a destination and the GPS receiver will calculate the path to the destination and guide the user to the destination with explicit turn data. The destination itself may have been the result of a search of the database based on an address. Among the handheld receivers released from Garmin only the GPS V can actually perform autorouting on the receiver itself. In order for autorouting to work the receiver needs a map of all of the roads that are available and the database used to produce the map requires much more data than just the connections between the turns. For example it needs to know if the street is a one way street, if left turns, or for that matter if any turns, are permitted at a given intersection, and much more very specific routing data. For this reason, a receiver that supports autorouting will not be able to autoroute without a suitable map database. In addition it needs to know additional data to make an informed opinion on what might be the best route to use when there are multiple choices. The speed limit for each segment of the road is an example of this kind of information. Freeway interchanges make the routing job even worse since the turn is often in a different direction than the final desired direction. Given the amount of complicated calculations it is pretty amazing that any handheld can do a decent job of computing a route automatically.

While only the GPS V can do autorouting on the unit itself any of the Garmin handheld receivers can provide navigation for a route generated externally and then downloaded to the unit via a suitable program. For example the Garmin Mapsource program can use the same maps that are downloaded to a mapping capable unit to perform an autoroute on the pc and then download both the maps and the resultant route to the unit for execution. Even units that do not support maps can use a downloaded route to guide the user to the destination. Third party programs such as Delorme Street Atlas can also autoroute on the pc and then download the resultant route to a Garmin receiver however only Mapsource can also download the exact map that was used to generate the route.

Getting the right Map

The GPS V comes with a copy of Mapsource and a set of maps called "City Select". These maps have been optimized for use on the small screen of the GPS V by limiting the length of some of the instructions so that they will fit on the screen. There are "City Select" maps available for the USA, Europe, Australia, and parts of Canada. "City Select" maps are made and licensed from NavTech (except the ones for Australia), which is a major map supplier that also supplies maps for most of the GPS systems sold as preinstalled in automobiles. As part of the purchase price the user in entitled to unlock one region from the cdrom. The rest of the cdrom can be unlocked for an extra fee. For Europe the City Select data includes not only major cities street level detail but also detail for the surrounding countryside and smaller towns. Unfortunately, at this time, this is not true for the USA where only major towns are covered with street level detail. A new release of full coverage for the USA has been announced.

In order for the GPS V to use a map for autorouting it must first be downloaded to the unit. There is 19Meg of memory available for the maps so the user may have to be choosy about which ones are loaded at any one point in time. Of course a different set can be downloaded as needed. Each download overwrites all of the previous maps so map loads are not incremental.

For the USA there is a another choice for maps available from Garmin called "MetroGuide USA". These maps also have the extra information needed for autorouting and cover the full USA with street level detail. In addition there are no unlock codes so the purchase includes the full cdrom set. These maps are made by and licensed from ETAK, a division of Tele Atlas, which is a competitor to NavTech. They are good maps but some have reported that they are not quite as up to date as some of the "City Select" maps in cities and do not have quite as much routing information. However, they can do a good job and provide the only available autorouting solution for many parts of the USA. It is possible to download a mix of "City Select" maps and MetroGuide maps so long as it is done in the same session.

Garmin has also released a set of maps for Europe called "MetroGuide Europe". This can be a bit confusing because these maps are from NavTech and are the same maps used in "City Select." They do not support autorouting on the GPS V but do support autorouting on the pc thus the results can be downloaded to he GPS V. They do not require regional unlock codes.

There is a brand new set of MetroGuide maps for Australia as well. These come from GME Electrophone, which is also the source of City Select maps. The MetroGuide maps are reported to have more detail in the outlying areas than the City Select product.

The GPS V can also use its basemap for autorouting thus it is possible to have a mix of maps that cover a single route. The GPS will automatically use the 'best' map(s) it has for the area that it is trying to route. It will choose "City Select" first, "MetroGuide USA" second, and finally the basemap. There is usually no need to have detailed maps of the entire journey in the unit. You might have loaded detailed maps for the start of the journey and the end of the journey and use the basemap for the road coverage in between.

It is possible to load any of the other maps available from Garmin into the GPS V as well, however they will not be used for autorouting. Of course the GPS V also supports manual routing, which is covered in the chapter on routes. Manual routing is used when the destination or route is off road. It may also be the best solution for long over the road routes between cities.

Performing an AutoRoute

Route creation begins by visiting the main menu (press the menu key twice) and then highlighting and selecting the route command. Select the "New" button to begin a new route. The following steps are used to create the route:

1. Select the type of route, "automatic" from the menu. (Creating a manual route ("offroad") is covered in the route chapter.)
2. Select the desired destination by category from the Find menu. All of the choices are available for autorouting. However some of the choices may have a final leg that is off street since the location may not be on a road.

3. Once the choice is on the screen select the "GOTO" entry and press enter.
4. A menu will appear that offers "Faster Time", "Shorter Distance", or "offroad". Select one of the first two for your autoroute. If the final destination is off road and you have not selected offroad the autoroute command will compute a route that takes you to the closest road perpendicular to the desired destination.

A small box will appear in the corner showing that the GPS is computing a route. Once the route is completed the unit will automatically begin navigating the route it just computed. A new turn by turn current route page will appear in the page rotation for the GPS.

It is also possible and perhaps simpler to begin a route from the find menu. Press the find button and proceed from step 2 above.

If the route is too complicated you may see a message saying "Too many vias for Road Navigation". This means that the route requires too many turns for the unit to calculate (more than 250). To recover from this error either compute the route manually (see below on using manual and automatic routes together) or choose an intermediate destination and retry the calculation.

Saving the route

It is not necessary to save the route before you use it. However if you want to save the route for future use it can be saved by going to the active current route page and selecting "save route" from the local menu. It will be named automatically and a small car icon will show to the left of the name on the route page to indicate that this is an automatically created route (manually created routes show a hiker).

Note that saved routes can be used in the future but cannot be reversed like a manual route can. This is because the turn by turn instructions is very dependent on lanes, freeway exits, one way streets, etc. It is possible to save two routes, one in each direction. To use a saved route, visit the route page from the main menu, select the route you want, and answer the popup window with "Yes" that you do want to navigate this route.

The route page is also the place that allows you to rename the route or delete it. Select the route and then press the menu button to see a list of options. Select the one you wish.

Creating an automatic route from a different starting point

Generally the GPS V will always expect that you want to create a route from your current location. However, for planning purposes, you may wish to generate a few routes ahead of time. To do this you can use the simulation mode on the GPS to move your current location to the desired starting point. Go to the satellite status screen, select Use Indoors (simulation mode), and then select a new location from the local menu. Once you have selected the desired starting point then perform a normal automatic route calculation, and then switch to the current route screen and save your new route.

Using an AutoRoute

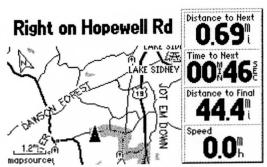

Figure 15 GPS V Map Screen

You will generally use the map page and perhaps the active route page when running a route. The map page shows different information when navigating than it does when just viewing the map while using the GPS for position. Both settings are customizable from the map page local menu. In addition separate settings can be obtained by switching the screen from landscape to portrait mode (press and hold the page key). The defaults setting, however, are likely to be the ones you want.

When you customize the navigation menu you can select the 4 entries as shown or only 3 entries with the upper one replaced with a double sized box that has a direction arrow in it. The large arrow provides visual direction and turn information including the approximate amount to expect from an upcoming turn similar to the arrows shown on the active route page. Even with the 4 box choice a smaller straight arrow is available.

The map page will show guidance text at the top of the screen (may be turned off if desired from the local menu) which provides essential information for the next turn. When not navigating this information shows the name of the next street or next freeway exit. Sometimes this is useful even when navigating. To see this data go to the active route page and select suspend routing (Do not select stop routing on the map page.) and the next road data will appear at the top of the map page. Select resume navigation to return to turn by turn instructions.

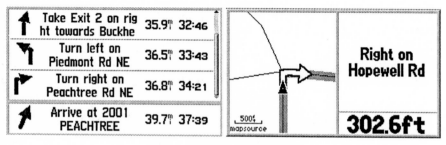

Figure 16 Active Route

The active route page, shown on the left in figure 16 displays the information on the next turn and the following 3 turns. It is updated dynamically as you travel the route. You can choose whether to show the projected arrival times or the times to go for each of these turns from a choice on the local menu. These times are computed using a proprietary algorithm but seems to be based mostly on the speed limits assigned to the various roads. It will not recompute them if you slow down for a turn or stop for a traffic light except to extend them for the amount of the stop.

Pressing enter while on this page will bring up a detailed turn map, shown on the right in figure 16 for the next active turn or if you select one of the other turns it will bring up the detailed active turn map for that turn. These maps always assume track up so that they can be interpreted as left or right turns. These are the same maps that will popup automatically as you approach each turn (if enabled) no matter what page you are currently viewing.

The local menu is also the place to temporarily suspend a route, resume the active route, recalculate the route, or force a detour calculation. Selecting a detour brings up a screen that allows you to mark the current road as impassable for a specified distance. The GPS will then calculate an alternate route that does not use this road. (This may require a U-turn.)

As you follow the route you will find that the GPS V will give you ample warning for each upcoming turn. It provides a beep early on to alert you to an upcoming turn and a second beep when you are almost on top of the turn. The first beep is a variable time based on your current speed (a feature also of the emap product). At night, if the backlight display has timed out, the backlight will turn on to alert you to the turn. You can set the maps to change scale as you near the turn but generally this is not as necessary as it might be on other units since the detailed pop up screen will provide a map for the turn itself.

If you get off route the GPS V will detect this condition and provide a message. It will say "Off Route" or if you have automatic recalculation enabled it will say "Off Route - recalculating" and will provide a new solution when the calculation is complete. If you continue off route without recalculating you will need to use the map of the highlighted route plus the arrow, if configured, to find your way back.

Recalculations

There are a number of times when you may wish to perform a recalculation of the route. For example, perhaps you got off route and need to find your way back. Or perhaps the road ahead is blocked and you need to detour around the area. Maybe the original route was been done just for over the road use and you now need a more accurate detailed final route. Each time the unit performs a recalculation it basically starts over from your current location and computes the full solution again. A detour is a special case in that you can specify not to use the current road for a certain distance. It does not just calculate a route back to the original route since there may be a better way from where you are and you are often closer to the destination when the recalculation is requested which may result in more detailed information. This is particularly true when the original route was calculated using a setting less than the "best route".

Via Points

A new firmware release to the GPS V has added support for via points. These points are useful to guide the autorouter along a specific path. When you set a via point the auto router will go through this point to reach the destination. Via points can be used when you want an intermediate stopping point on the route or to correct a route that is not going the direction you wish. They can be added after the route is generated to correct a part of the route and they can be removed if you decide the original route is better. The active route page contains a local menu entry to add a via point to an existing active a route or to remove one if there is one already present. After adding or removing a via point the route will be recalculated.

The new via point functionality can also be used with manual routes. In this case, the user would define a manual route containing the major turn points. The Route List page contains a menu item called follow roads. Selecting this entry will cause the autorouter to define a path that follows the roads and goes through each of the waypoints in the route.

Setup Preferences

There are several setup items that can be used to enhance and control the autorouting capability.

Map Page Setup

On the map page you will want to select the lock to road feature. You will also need to enable the maps your wish to use using the Mapsource Info menu item. It is possible that the routing and the display are using different maps. For example if you routed an area on the basemap but have the topo maps enabled you might be looking at an area on the topo map and using the basemap for autorouting. This can cause discrepancies between what you are seeing and what the guidance being offered by the autorouter.

Main Menu Setup

The main menu setup choice provides two entries for automatic routing use, guidance and routing.

Guidance setup determines how the GPS behaves when using the active route. There are two entries on the page.

- Off route recalculation can be set to automatic or to prompt you if a recalculation is needed. If you leave the route the GPS v will detect this fact and can recalculate a new route based on your current location if you wish. You may prefer to just return to the old route since you left the route only temporarily to goto a restaurant or a gas station perhaps.
- Next Turn automatic popups can be turned on or off. When on these detailed turn directions will appear automatically when you are about 10 seconds away from the turn. When off you can still display this information manual from the current turn page by pressing the enter key.
- Routing setup controls the way the routing works. There are several setting on this page.
- Routing preferences can choose between three preferences.
- Faster time - The preference uses database information about the speed limits on certain roads to determine the best and quickest route to the destination. There can be distinct differences in the calculations based on the maps you use. Some map databases may not have the latest or detailed information about the type of roads on the maps and may not take advantage of a faster road for this reason. Generally City Select has the best data for this use.
- Shorter Distance - this setting ignores the speed limits in determining the route.

- Off Road - This setting disables the autorouting completely and computes a straight line to the destination for all gotos.
- Ask my preference - this setting will cause the GPS to prompt you for your preference each time a route is to be generated.
- Calculate Method - This setting determines the algorithm used by the autorouter. You can choose between best route, better route, quick route, and quickest route. These 4 choices determine how much time the unit will spend trying to figure out the solution, how many alternatives it will consider, and how much data it will factor into the computation. All of the choices will result in a solution but the quickest will generally report only the first solution the unit found while the best will try all possible alternatives and all available database data to determine the presented solution.
- The Calculates routes for setting determines the type of vehicle the router should expect. Often the type of vehicle determines the best route since a commercial truck, or a bicycle may not be able to go on certain roads or may experience lane restrictions. Speed of the roads doesn't matter much to a pedestrian and emergency vehicles don't care about certain rules of road usage such as 'no left turn'.
- The entries at the bottom of the form allow some control over specific types of routes but cannot be guaranteed to be followed by the router since they could prevent a solution. Selections include "no U turns", no "Toll roads", and "No highways".

Using Manual Routes and Autoroutes Together

There are lots of reasons you may need or want to use a manual route in the GPS V. Here are a few examples:

- you used a 3rd party program to calculate the route for your vacation and wish to download this into the GPS v.

- Perhaps you have a scenic route in mind and would like to dictate the intermediate points along the way. Maybe someone sent you specific instructions to find their house, which may be better than the route the unit might compute.
- There are cases where there is a locked gate across the road that the map database didn't know about.
- You may have traveled to a location and kicked off the trackback feature to return using the saved tracklog.
- And finally, the trip you are planning is too complicated for the router or you don't have the correct maps downloaded into the unit at the time you setup the original plans. This can easily happen on a vacation where you intend to load the next days maps into the unit each evening.

However, it can happen that while traveling using a manual route you encounter a condition where you need an automatic route. For example you miss a turn and need a calculation to get you back on route, or perhaps there is a detour that you didn't know about. While the route is still active press and hold the find key. This will bring up the active route screen and you can select the desired destination. When the waypoint screen appears select the goto button. This will cause a calculation of a route to that point as a destination that you can then follow to get back on track. Note that after the automatic calculation you will need to manually re-enable the original route if you still need it.

Chapter 10
Working With Navigation

While a GPS would be a great device if it merely told you where you were, in practice it will tell you a whole lot more. This chapter will focus on its ability to help guide you to your destination. This capability can be useful and might save your life some day. Navigation has been traditionally performed using a compass and a map along with your eyes to help you relate where you are to where their location is displayed on a map. Off road hikers and pilots often supplement this information with an altimeter that can aid in finding your location on a topographic map. In some cases a GPS can actually replace both the map and the compass, and to lesser degree the altimeter, however it is a good idea to have a compass and a map as a back up along with the skills necessary to use them.

Navigation Basics

Before starting to learn about the GPS navigation capabilities it would be helpful to spend some time understanding the basic terms used to describe navigation. There are three basic kinds of navigation in use today, land, sea and in the air. All three have their own terms for certain things so you may find the same information labeled different ways depending on which form of navigation is being discussed. Garmin has units specifically designed to help in each of these operating areas but has divided their product line into land/marine and aviation. This, however, does not mean you can't use one GPS in another environment but some features are specific to a particular use. Some folks would like under the sea navigation as well. Unfortunately a few millimeters of solid water makes a GPS useless so, unless you can float an antenna, a GPS will not help for this type of navigation.

To help in understanding the basics of navigation please refer to the diagram in figure 17 below as needed:

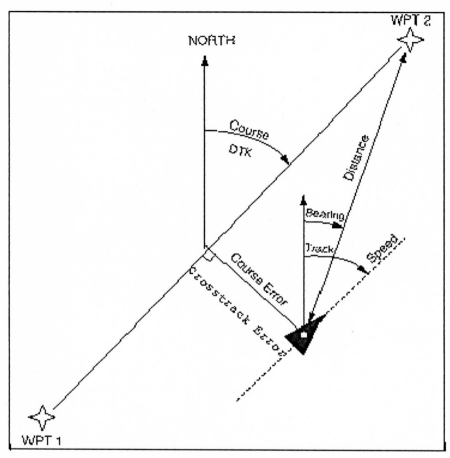

Figure 17 Navigation Basics

When you want to go somewhere you probably have a starting point and a destination in mind. If you were to go directly to the destination you would just set out and head directly for it. In this case you would be traveling on a particular **track** or **course**. If you were to sight your compass towards the destination you would call it a **bearing**. Now if you weren't able to go directly toward the destination because of an obstacle or perhaps the road doesn't go that way your track would change but your *desired track*, 'DTK', would stay the same. As your track takes you out of the way of reaching your goal the bearing will be changed. By the way, for land travel, the term heading and bearing are almost interchangeable but for boats and planes they mean two different things. Bearing is the direction to your target and **heading** is the direction you are trying to go, the way your boat or plane is pointed. Often, because of currents or winds, you will need to head in a different direction than your actual goal to offset these external forces.

If you could see your destination from the beginning you probably wouldn't need much navigation help in order to get there. Most of the time, however, you can't see the location but may have located it on a map. Using traditional techniques you might draw a line on the map from your current location to the destination and then use a protractor to figure out the bearing. You would then use a compass to keep your track as close to the bearing as you could. Drifting off course could cause you to get the map out and figure out where you are and compute a new bearing. If a straight line is not possible then you might build a route that took you somewhat out of the way to make the trip itself easier. Then you would have a succession of bearings and DTK's called legs that would, together, allow you to reach your goal. This collection of legs is called a route and the individual points where the turns are done are called waypoints on the route.

If you were in a plane or on a boat and had some cross wind you might drift off of the desired track even though your track as indicated on the compass never changed. The distance you've drifted is called **crosstrack error** or course error. This may or may not be important depending on your circumstances. For example if you were sailing in a channel drifting off course might cause you to run aground which could ruin your whole day. In this case you would want your navigation equipment to alert you to the fact that you are drifting and provide some guidance back to the optimum course. A GPS might indicate this kind of error in a number of ways. The bearing and track are updated every second or so and if the bearing begins to differ from the track then you are clearly moving off your course. This relationship between track and bearing is important enough that Garmin considers it to be two of the four main pieces of navigation information. The other two are your current ground speed and the distance to the destination. Garmin places these four pieces of information on the map screen and on all navigation screens.

While this information may be enough for many folks, some will want more details or, as in the case of the guy in the boat, more information about getting back to the original course. (Following the bearing will get you to the destination but may not keep you on the original course.) Some Garmin GPS's provide the course to steer, **CTS**. This is an actual compass heading to turn your craft toward in order to get back on course. By the way this information can be misleading if you are fairly close to your destination. The computation looks at the angle of the current track and the desired track and computes the most efficient course to get back on track. This might actually overshoot the waypoint so always check it against the bearing. If you prefer a relative direction then you might like **TRN**, turn, which provides information on how much you need to turn to get to your destination.

Other interesting data includes **ETE**, estimated time enroute, and **ETA**, estimated time of arrival. To compute these numbers you need to use an estimate of your speed. Garmin calculates the speed vector that is actually in the direction of the destination and calls this **VMG**, velocity made good, based on your current speed and direction. The handhelds covered in this manual (except for the etrex and emap, see below) all use the current VMG as the effective speed for calculating ETE and ETA. VMG is a pretty good indicator of your effective speed if you are moving at a steady speed over the ground and are approximately headed in the right direction. It is also accurate if you are in a sailboat tacking toward an up-wind mark by zigzagging back and forth. Similarly a car traveling on city streets may have to do the same thing but the destination may not be directly "up wind". Either take the VMG on the closest tack in the direction of the destination or better yet average the two to arrive at the most probable ETA.

The etrex and emap also use VMG to estimate ETE and ETA, however the way it is measured differs on these units. Instead of a current VMG they use a previous sustained velocity to calculate VMG. Changes in speed or direction will not change the VMG until the magnitude and direction settles down to a new constant speed. In addition no change in calculation of VMG is done when you are stopped. Under these condition the old VMG will be used to update the ETE and ETA values. Unfortunately there is no way to display the VMG that the unit is currently using, however you can force it to compute a new one. To do this, temporarily stop navigation from the local menu and then start it again.

Garmin generally provides two navigation screens for use in guiding the user to the destination. These are called the Compass (or pointer) screen and the highway screen. These are illustrated below in figure 18 for the G-12 family.

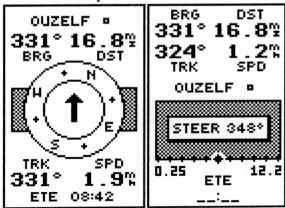

Figure 18 Nav Screens

Generally a user will prefer on or the other. Surprisingly, based on the name, land navigation is usually done with the compass screen and water and air navigation is often done on the highway screen. Folks with a real highway don't need a GPS to keep them on the road but folks without a real highway can benefit from a virtual highway that keeps them centered within their prescribed course. This is exactly what the virtual highway screen does by featuring a visual display that illustrates any crosstrack error that may be present. On many Garmin units only one of the two pages shows up in the screen rotation. (The G-III family has both in the rotation and the etrex only has a compass screen, which it calls a pointer screen.) To select which one will appear in the rotation press the enter key without any field highlighted. You will be given a menu that can select which screen you want. Press enter again to switch.

Each screen supplies basic graphic and textual data. By default non-mapping receivers also show bearing, distance, speed and track while mapping receivers seem to have replaced track with ETE. While mapping receiver provide more customization of the data displayed they will only display 4 things while non-mapping receivers will often display more.

The emap does not have a dedicated navigation screen but instead it changes the top area of the map screen for navigation use. Instead of a compass display it just shows an arrow pointing in the direction to the next waypoint. It also shows your current speed, the distance to go to arrive at the waypoint, and the estimated time to arrive. If you are stopped the pointer arrow is replaced with a compass bearing. This can be displayed as a number in degrees or as text using the 8 points of a compass. The destination name is not displayed on this screen but you can reach it easily by pressing and holding the find button. It will switch to the waypoint page if you are doing a "goto" or to the active route page if you are navigating a route with the current destination indicated. If you use automatic zoom then the destination will be shown on the map itself.

Working with the Compass Screen

Figure 19 Compass Screen

The Compass screen, G-III version shown above, is called that because of the large circular area that dominates the screen. It looks like a compass but it is not. Garmin units do not have a compass built in but simulate a compass using Doppler position information to deduce your direction of travel. In an attempt to avoid this confusion the latest Garmin offering, the etrex, calls this a pointer page focusing on the ability of this screen to provide a pointer to the destination waypoint. If you visit this page when you are not actively navigating to a destination the pointer in the center of the screen will be missing but the circles with still rotate to give you a simulated compass with your current direction at the top of the screen. In addition your current speed and track will be shown (or other data you have chosen on the G-III family). These two entries will be changed to underscores if you ever lose your satellite lock.

The easiest way to use this screen is to use it with the goto key[*] and an existing waypoint. Press goto, select a waypoint, press enter and the navigation screen will show you the waypoint name at the top of the screen, the direction to the waypoint, and will begin guiding you to this destination. The GPS will compute both the bearing and the great circle distance to the waypoint. For many folks this is all of the navigation they will ever need. Many units also provide for an emergency man overboard capability. Press the goto key twice, press enter and you will generate a waypoint at your current location called MOB and start the navigation screen with information to guide you to this point you just left to help you recover the person who just fell overboard. You may be able to think up many other uses for this capability.

Due to the fact that track is a computed direction based on movement it can't be depended on when you are stopped. The GPS tries to pin this value to the last computed value while moving but the GPS can become confused about movement and basically makes this number meaningless. However bearing works even when you are stopped. You may need a compass to actually use the bearing data but at least the value is correct.

Once begun the GPS will continue to navigate toward the destination waypoint until canceled. The cancel goto command is one of the menu selection items reached from the goto key itself. The easiest way to reach it may be to go 'up' from the menu, which will wrap around and select the entry on the bottom. Or simply continue to press the down arrow key until the cancel command is reached. On some machines pressing the right arrow key will skip all of the waypoint names and jump directly to the cancel goto command.

On the emap and etrex you can use the local menu item, stop navigation. This command has the added benefit that it remembers the previous navigation setting so you can re-enable it using the start navigation item if you wish. This can be useful to remove the

---

[*] Note: the etrex and emap work differently here. There is no goto key so you first select the waypoint you want from the waypoint list and then one of the review choices is 'goto'. Selecting goto from this list will initiate the goto as described above. The goto can be canceled from the local menu available on either the pointer screen or the map screen.

navigation goto line temporarily from the map screen. Turning off the unit with navigation stopped will end the navigation session completely.

Some older GPS units and the G-III pilot also have a graphic course deviation indicator, CDI on this screen. This is an indicator for crosstrack error. However, most units reserve this information for the highway screen. Its use will be covered in the description of that screen. If you want to view the crosstrack error and you have a selectable field you can select XTE for a numeric readout of this error. Because the etrex is intended for land navigation it does not display or compute XTE. The map screen on the etrex has a setting for the goto graphic line to show bearing to the destination or just a static line that shows the desired track from the start of the goto. This can be used visually as a crude cdi indicator.

In addition to the fields and information described above your unit may offer to display other optional entries (older multiplex receivers do not offer any options but always display track, speed, bearing and distance). On the etrex you can always view the ETE and distance and can press the up/down keys to select bearing, track, speed, and some other entries not directly related to navigating. The III family can customize the display to whatever degree you wish using the local menu to select among navigation data and other entries as well. Other units always display speed, track, bearing, and distance (the golden four) and you can highlight the bottom entry and press enter to select among ETE, ETA, XTE, VMG, CTS, and TRN.

Some folks like to leave the unit on the compass page whether or not they are currently navigating. This is because it has the largest text display of any of the screens and can be seen from a distance. On the III family this can be enhanced by selecting the small compass from the local menu. When the small compass is displayed there are two very large customizable entries to view. The etrex and the III family can display information that is useful even when you are not navigating. The etrex, which has no position screen, will also display essential position data using the up/down keys such as altitude, latitude/longitude, max speed, average speed, trip distance, trip time, and even sunrise and sunset times. Max speed and trip information may be reset from the local menu. The III family can display any of 29 or more different things on these pages. You can basically configure any of the III screens to display any data with just a few exceptions.

The HSI display

The compass screen differs considerably on the III Pilot. It offers a display that is more familiar to pilots called the HSI display, horizontal situation indicator. The HSI display has much more information available than the display on the other units. Consider the figure shown below:

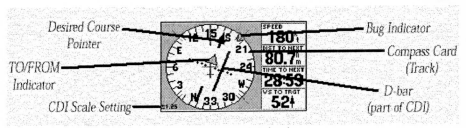

Figure 20 HSI Display

As can be seen the compass image has changed from just a simple display that looks like a compass. In addition to the compass rotating to indicate your track over the ground and a moving arrow, you will notice the CDI display and a small triangle showing whether you are really headed toward or away from the desired waypoint. This will provide a visual indication as you fly over the waypoint. The current CDI scale is shown in the lower left corner and can be changed using the zoom keys. The D-bar and CDI information will be covered in the Highway section below. Differing from other GPS navigation units the III Pilot will let you override the course from the local menu. You can chose 'OBS and Hold' to force a course of your choice, selected with the left and right arrow keys. The arrow and CDI will then provide navigation on your selected course. The actual course direction you selected will show up as text just below the center of the display. Once set, this heading will not change until you change it. You can release the hold which will then keep the setting until the waypoint in reached. Since the GPS always computes information relative to ground this can provide heading compensation capability.

Vertical Navigation

Some units can provide vertical navigation. These include the III Pilot and the etrex Legend, Venture, and Vista.

The III Pilot can also provide vertical navigation using the HSI display. You must first go to the main menu and define a vertical profile for your glide descent and the final target altitude. Once this is done the unit will automatically generate a message when you approach the vertical navigation profile. You can then turn vertical navigation on from the local menu on the HSI screen. A horizontal bar will appear on top of the display and indicate your glide slope. It will guide your descent and will disappear 500 feet before your target altitude. Remember that satellite position errors can cause variations in the GPS calculated altitude so this data should be treated as information only and not used to provide a precision approach. GPS receivers normally optimize satellites for horizontal position accuracy.

Vertical navigation on the Vista is provided with vertical speed data and glide slope. Glide slope provides a ratio of the amount of fall in altitude vs. the distance traveled. It is computed only when navigation is in progress. There are two selectable fields, one shows the glide slope to the next waypoint in a route while the second shows the glide slope to the final destination. It would be nice to use this to identify the time to start a descent but unfortunately it will not report a glide slope when traveling level so you must actually start a descent before it will compute the glide slope. It cannot be used to compute an ascent. Of course, the destination waypoint must have altitude stored in the waypoint for this function to work properly. Otherwise it assumes sea level for the calculation.

Working with the Highway Screen

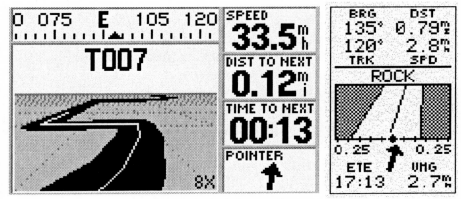

Figure 21 Highway Screen

The highway screen is really oriented around following a prescribed course. Pressing the Goto key as described in the above compass discussion not only set a destination but also set a beginning point as your current location the second you pressed the button. The line drawn between these two points is your desired course to reach the waypoint in the minimum time and distance. Use this screen when it is important to maintain a track that matches the "desired track" or "course". If you need to reset the course you can hit the goto command again. It will highlight the same waypoint you used last time and pressing enter will reset the starting point to the new present location. Note that the etrex and emap do not have a highway screen.

While all of the highway screens in the various units look similar (the G-III version is shown above on the left) there is a wide variation in how they actually perform in use. The dominant feature is, of course, the highway itself. This shows a three dimensional view of a virtual highway that points from your current location to the destination waypoint. In the case of a route the waypoint is the current intermediate destination. Perspective is used to give the 3D effect and if you ever head the wrong direction you will find yourself at the narrow end of the road! If the highway is centered in the display then you are heading for the waypoint and are on the centerline of the course established when the goto, or route, was initiated. If you drift off the course then the highway will drift left or right to indicate how you should steer to get back to the road. If you head off in the wrong direction the road will turn to indicate that you need to turn to get back to the road. This basic visual information becomes quite intuitive with use and is the same on all of the highway screens.

Somewhat surprisingly the older single channel units provide the most data on this screen as shown above in figure 20 to the right. This coupled with the good performance a single channel unit can support under the open sky that usually accompanies water has led to a long life for these units among boaters. These older units show the four main pieces of navigation information, bearing, distance, speed and track, the name of the destination waypoint, a cdi scale, the ETE, the VMG and an arrow pointing to the destination in addition to the road drawing. Newer non-mapping units show the same thing except that there is no bearing arrow and you must choose ETE, VMG, XTE, TRN, CTS, or ETA. You get a wider choice but you can only choose one to display. This can be a different one than you chose for the compass screen. The inclusion of the cdi scale shows a strong tie to marine navigation.

The mapping units show their tie to land navigation by leaving off the cdi scale and making the road screen a snazzy display with good visual impact but less definitive information. Note that the III pilot uses the HSI screen for cdi so it doesn't need this information here. Mapping receivers can use the zoom keys to zoom the highway and are able to show route destination beyond the current target waypoint. This can be used to predict the direction of the follow-on waypoints. You can even use a local menu item to turn on signposts on the highway screen that indicate the names of the route waypoints. The III pilot will even show airport waypoints that may be close to the "road". Similar to the compass screen you can customize the 4 numeric display fields. You can show different things than are shown on the compass screen and even select different things between the horizontal display and the vertical display if your unit has this feature. An edge view of a compass display is also shown to indicate your track.

Using CDI

The Course Deviation Indicator, CDI, is key to maintaining your position as you traverse from one waypoint to another in water or air.

The numbers provide a maximum deviation for the display and this scale can be changed. On many unit you will need to go to the main menu and select navigation setup but on some units with dedicated zoom keys you can use these keys to change the scale. Changing the scale visually changes the width of the road so it looks similar to a zoom function. A larger scale permits more deviation before the CDI changes. If you drift more than 1/5 the distance on the CDI scale the maximum scale indicator on the side of the drift changes to indicate the exact amount of drift. If you drift so far as to cause the road to totally disappear from the screen it will be replaced with a course to steer, CTS, message telling you how to most efficiently return to the desired track. When you finally arrive close to your waypoint a line drawn across the highway will indicate the distance. If you are following a route and this is not the final destination the highway will switch to point to the new destination when you reach the line. Some units also have an alarm that can be set from the main menu, setup screen. Setting this alarm will permit a visual and perhaps an audible alarm to be sounded if you exceed your predefined deviation from the desired course. Other alarms can be set to indicate that you are nearing arrival at your destination. These alarms will work even if you are not displaying a navigation screen. A visual alarm displays a message on the screen at the alarm period and at night, if you have the light turned on and it has timed out, it will turn on the light to display the message. If you are using external power then the lamp timeout is usually disabled on Garmin units so you will only get the message on the screen. Units with audible alarms will beep when the alarm condition has been met.

Other Screens and Information

The map screen has been described in detail in the chapter on Garmin screens but its use as a navigation screen should not be overlooked. Not only can it provide the 4 pieces of essential navigation data textually, but the map display itself has a high value for visual navigation. Recall that the emap provides all of its navigation capability on this screen. Creating and displaying a route on this page can give a sense of overall direction and on many units this is the most convenient place to view and anticipate upcoming changes in course. Setting the screen to desired track, DTK, up or track up can provide a heads up indication of drifting and upcoming hazards you may have previously marked. The overhead view of the route as provided in this display is intuitive and valuable even without an underlying map. Having a few extra waypoints can provide orientation. Units with internal databases can provide navigation aids via labeled icons on this page. A visual indication of cdi can also be discerned from observing this page.

If your unit provides maps built in then the map page may provide even more visual navigation information. If you have a unit that can upload maps you can upload topo maps and even include altitude information you read from the elevation lines on the map as part of your navigation decision. For the III pilot the map page can provide essential Jeppesen data marking VOR boundaries and other essential navigation aids. See the database chapter for specific details on what navigation aids are available in the various databases.

Mapping receivers also have text fields that can show ETE at the destination waypoint and ETA at the destination waypoint for a route. Other units cannot display this directly on a navigation screen but can get at the data by going to the active route screen and scrolling to the last waypoint in the route. This information along with the desired track and cumulative distance is available for every waypoint on the route. Even the older units have this entry as a customizable field on the route page.

The active route page should not be overlooked as a source of navigation information. On the emap this screen is the only place that you can see the ETA values. This information is valuable on any of the handhelds. If no waypoint is selected or the last waypoint is selected you will see the ETA at the destination; while selecting a waypoint will give ETA to that point. You can quickly reach the active route screen on the emap by pressing and holding the find key whenever a route is active.

All of the units can indicate navigation warnings. Units with sound capability can couple the visual message with a sound if desired. Messages include upcoming waypoints in a route, warnings about hazards, and even warnings that your the upcoming turn is too tight if you are traveling fast enough to be assumed to be an airplane. Turn smoothing starts to appear at 75 mph. The upcoming waypoints message can be automatically generated based on your estimated time of arrival assuming you maintain your current speed or in some units it can be set at a specific distance. The automatic arrival alarm is fixed at one minute on all units except the etrex and emap. On the etrex is set to 15 seconds and the emap has a variable time based on your speed. In some cases you have control over these warning messages and in some units you do not. If you can change them there will be an alarm entry on the main menu page perhaps as a submenu under system.

Some of the units intended for marine use have proximity alarms. These can be set to alert you if you get to close to a hazard. Each alarm can be set to a different distance. If you have proximity alarms you will find them on the main menu.

Tips and Tricks

Here are a few tips to help with GPS navigation:

- If you are using your GPS with a map and the map is oriented to true north then setting the GPS display angles relative to true north will make things easier. If you map is oriented to grid north (most often UTM or Military maps) and your GPS has provisions for grid north then use it, but you will only be off a maximum of about 3 degrees at the corners if you don't have this option. Be sure the GPS is set to match the map datum.
- If you need to use an actual compass, for example to steer by, along with your GPS then you will probably need to set your GPS to magnetic north to agree with the compass. A few land compasses can compensate for the difference so you can use true north for both but most nautical compasses cannot.
- The MOB feature can be used for a quick trip distance meter. For example if you see a sign that says exit 3 miles ahead you can set mob and then notice that the distance to mob will be steadily increasing until you get to the exit you want at 3 miles.
- While the emap does not have an MOB function you can use the measure distance command as a workaround.

Select measure distance from the local menu and the display will temporarily provide much of the same benefits as an MOB waypoint.

- Do not expect any navigation compass displays to be accurate when you are not moving. The bearing will still be ok but you will need an actual compass to take a sighting on that bearing. Some units do have an actual compass built-in, which can be turned on when you are stopped.

- If you become lost or disoriented then the GPS may be able to get you home. Check the nearest waypoint screen for any waypoints you may have set to places close by. You can use trackback even if you only turned you unit on to take occasional fixes so long as the track log was enabled.

- A standard altimeter is often used with a paper topo map and compass as an aid to finding your location. You can locate the trail on the map and by looking around or taking sightings and you can often use the elevation marks to pinpoint your exact location. You cannot use a GPS altitude reading to do this as it is not accurate enough unless you are using dpgs. Unfortunately a standard altimeter will drift over time or because of weather changes. A GPS location fix can often help to pinpoint your location on a topo map, even one inside the GPS, which can then be used to calibrate the altimeter!

- Suppose you notice something off in the distance and would like to go find it. For example, a model rocket that you launched or perhaps an animal you spotted off in the distance. Take a bearing using a compass and then project a waypoint from the bearing and a distance that is further than you expect the object to be. Then set a goto to the projected point. Make sure your map page is displaying a course line to the waypoint, which is standard on most Garmin GPS receivers, an option on the etrex, and not available on the emap. Now you have a visual line that can be used to guide your search efforts. Once you find the object you will be able to easily find the actual distance.

- Navigation is often used in conjunction with maps however a problem appears if the map does not contain a grid or other reference data. One way around this problem is to locate on point on the map, such a trailhead, and set a waypoint in the GPS that matches this point on the map. Setting a goto on the GPS at any time will tell you the distance and bearing back to this point. If the map has a reasonable scale and is otherwise accurate you can use this distance setting and bearing to locate your position by computing a back vector from the saved waypoint. A protractor can be used to measure the angle and a ruler or dividers can be used to measure the distance.

- One of the common uses of a GPS is to leave it off most of the time and then take occasional fixes to locate a position on a paper map. The idea is to turn the GPS on in a clearing and let it find the lat/lon and look this up on a map. Used in this way a set of batteries in a GPS can last a long time. Most GPS receivers will automatically find your fix and then switch to the position page to display your location. It is a good idea to wait a minute more to allow the position to settle and to permit a 3D solution. The etrex and emap do not automatically switch to a screen that displays this data so, on these units, it is often easier to press and hold the enter key to bring up the waypoint screen. You can then check the location and if you don't want to save the waypoint just cancel it at this point (esc on emap, page on etrex)

Most of all, have fun using your GPS for navigation. Just be careful that you don't trust it too much. As with any instrument you need to know the limits.

Chapter 11
Working with Simulation Mode

Many new users of Garmin receivers have questions about simulation mode on their new units. Once they have worked through the getting started portion of the manual they wonder if they would ever want to use simulation mode again. This chapter explains what can be done with this useful feature.

Note that this mode is not always called simulation mode on the different models. On the etrex it is called "demo mode" or "GPS off" while the emap calls it "Use Indoors" on the local menu or "GPS off" on the setup menu.

On most units you can turn on simulation mode from the main menu. This is done by selecting "System Setup" and then toggling the mode entry from "normal" or "battery/power save" to "simulation" mode. Some units have "system setup" as a submenu choice of "setup". On the G-III family and some etrex models this is also available from the local menu on the satellite status page. This information is only remembered for this power on cycle and the unit will revert to its previous mode automatically after you turn it off. Of course you can also modify this mode manually at any time during this session.

Training Purposes

Certainly one of the first uses for simulation mode is to learn how to use the unit. Many of the Garmin manuals use this mode to demonstrate and introduce the main features of the product. In addition a store clerk may use simulation mode to demonstrate the product although you may find some of them who don't even know how to turn it on. To highlight this use the etrex unit calls this mode 'demo mode'. Demo mode differs from standard simulation mode in that a fixed speed is set (20 mph) which causes the GPS to appear to move. This can be used to follow a route.

Upon entering simulation mode you will find that the GPS seems to have a lock on satellites and everything works similarly to the way it works when you actually have a fix. The satellite status page shows that you have a lock on the satellite graphic display and indicates that you are in Simulation mode in the status text field. The rest of the fields in the unit do not show that you are in simulation mode but may have extra commands or other information and may perform differently. For example, the units with object oriented commands add selection capability for speed and track settings so that you can change them on any screen at any time to permit simulating actual movement while using the GPS. Menu oriented units set this information on the original simulation mode setup screen so if you wish to change it you need to revisit the setup screen. The etrex unit has a fixed setting for speed and track which in consistent with its being called a "demo" mode rather than a full "simulation". Some etrex units also have a "GPS off" mode, which is similar to demo mode except the speed is set to zero.

The emap has a simulation mode as of the 2.70 release of the firmware. Selecting "Use Indoors" from the map page local memory or "GPS OFF" from the main setup menu will include simulation mode by enabling the GPS screen. In previous versions this menu entry was disabled.

- Start the GPS and turn on simulation mode using on of the above methods.
- Select the GPS status screen
- You can now use the up/down arrow keys to set speed and the left/right arrow keys to set the direction. Observe the direction by watching a small circle at the edge of the large status circle. It starts out at north.
- Once the speed and direction are set you can leave the screen with escape and use the unit similar to other Garmin receivers in simulation mode.
- You can return to the GPS status screen to make adjusts in speed or direction.
- Turn off the unit or select use outdoors to go back to regular navigation.

You can practice navigation as well as familiarizing yourself with other features. The goto command can initiate a simulated goto with

176

one of the waypoints in your unit. You can add speed to actually "travel" to the waypoint. On older multiplex and G-12 family receivers the object-oriented interface will permit you to select speed on the position page or either navigation page and enter the speed you wish. You can also enter the direction by selecting and changing the track display. On the III family you can change to the compass/HSI display page and use the up/down arrows to change speed. The effect of cdi can be explored on the HSI or Highway page using the left/right arrow keys.

Waypoint and Route Maintenance

One of the uses of simulation mode is to allow for route and waypoint maintenance. The emap considers this use important enough to call simulation mode "inside mode" on that model. While you can perform this maintenance on a unit that is trying to track satellites you may get "poor GPS coverage" because you are using it indoors or other annoying messages. Setting simulation mode avoids these messages and as a bonus saves about 1/2 on the battery consumption since in this mode all of the power is removed from the receiver circuitry.

Use this mode when you are downloading or uploading data to and from your pc or when you are just reviewing and fixing names in a waypoint list or building new waypoints or routes from map data. This is also a good mode to use when you need to collapse a few duplicate waypoints that appear due to using the backtrack command. Decide which waypoint you want to keep and then use the 'change' command in the local menu for route editing to change to the waypoint you wish to use. You will need to delete the extra waypoints separately.

On the older multiplex units and the G12 family you can change the location directly on the position page just by selecting it and changing the numbers. This has a number of useful uses. For example you can set the numbers you wish, hit the mark key, and enter the newly created waypoint directly into a route. The route number is remembered from entry to entry so you can build up a route from a map very rapidly using the technique. Another use is to set the current location for viewing purposes on the map page. Since Garmin receivers only show the icons for the 9 closest waypoints to your current location this method of changing your current location can be used to view and perform waypoint maintenance on waypoints that are at far distances from your present actual position. Note that this change will not cause problems when you power off the unit. It will forget the simulation position and remember the previous real position.

The G-III family of units will not let you change your current simulation position in this way. If the new location is not too far you can use the simulator to "drive" there by setting a high speed and track direction. Otherwise there is a neat trick to move the current location. Bring up simulation mode and then simulate a re-initialization of the unit. After selecting this from the status page you can change the location using the map to position the cursor. Once accomplished you will find the simulation mode gets a simulated lock at the new location amazingly fast! At this point the 9 nearest waypoints will be computed for the new location.

Using your Garmin as a Calculator

Your GPS is also quite useful as a navigation calculator. All of the GPS units can perform graphical calculations and a few even have a regular calculator built in. The use of the regular calculator is covered in the miscellaneous chapter. Using simulation mode for this is not required but is recommended to save batteries and when indoors. While most units won't add and subtract all GPS units are certainly capable of doing several other navigation-related tasks.

A GPS can be used calculate the great circle distance between two points. Just enter the points of interest and then use the waypoint page to display the distance and bearing. Since your current position is the waypoint named _____ (6 underscores) you can also calculate the distance from your current location. To be honest it is even more accurate than a great circle distance since it is really the great ellipsoid distance that is being computed.

Your GPS supports more than 100 datums. Each waypoint is stored internally using WGS-84 but you can translate it easily to match whatever map datum you may need. Just select the map datum of your choice and the GPS will compute and display the new coordinates referenced to that datum. Similarly you can translate from one grid system to another. For example lat/lon can be displayed in degrees and decimal parts of a degree but the unit can just as easily translate those numbers and display in degrees, minutes, and seconds. One really useful conversion is from lat/lon to UTM coordinates.

Similarly you can translate distance from English to metric and back again simply by changing displayed units. You can also convert angular measurements the same way. The current defined magnetic declination will be shown on the auto-mag setup screen. This is done with a table that is stored inside the unit.

Navigation in Simulation Mode

Strange as it may sound there are some navigation features on some of the Garmin handhelds that can be used in simulation mode. In particular any of the newer units support a screen that shows the sun and moon positions. This can be used as a crude compass to determine North and can be quite effective under conditions where you can't get a fix or need to conserve batteries. For rough estimates hold the unit in front of you with North straight ahead and then rotate you body until the sun or moon in the display is approximately where it really is in the sky. You are now facing North. For more accuracy lay the unit on the ground and use a straw placed vertically along the edge of the unit to cast a shadow over the face. Align the unit so that the edge of the shadow from the straw splits the sun and hits the dot in the center of the circle. Now the front of the unit is facing North. Note that this is true north.

The Vista and Summit contain a built in compass and altimeter. These are certainly useful for navigation use even without the standard GPS features. Note that "demo mode" cannot be used for this since the compass is disabled in this mode but "GPS off" will work fine. If you have a topo map available in the unit you can use the altimeter to help determine your location relative to the map features and topo data. Then you can use the "New Location" setting on the status screen to move your apparent position to the location on the map that you have determined. When this is done the waypoints you have saved and the maps that are loaded can be used to plan your trip or guide you out of an area where there is no GPS coverage. Of course, the compass and altimeter are useful with a paper map as well.

Other Purposes

One favorite use for simulation mode is to playback a route you have saved. This can be overlayed on a map on a pc by activating a route and providing some speed to the GPS. This will cause it to run the route automatically turning at each waypoint and traveling on to the next. Since all functions are enabled in simulation mode you can turn on the NMEA interface and the output of the simulation run will look exactly like you were really traversing the route. This can be used to display your route on a map or to debug a mapping programs GPS interface.

A novel use, submitted by a reader, is to use the GPS as a timer. If you set a waypoint as a destination and provide some speed you can watch the screen and use the screen predictions as a timer. If you have an audible alarm feature on your GPS it will even alert you when the cake is done.

Chapter 12
Working with Databases

This chapter covers the use of databases in Garmin receivers. Only some Garmin receivers actually support and include databases so this chapter is specific to those units. Databases are of two types. There is text information that can be displayed to the user and there is graphical data that can be displayed on the map page. Generally there is also search capability to locate the information you wish. Note that Garmin receivers that support maps do not contain the kind of maps that you might be used to when comparing to traditional paper maps. Instead they contain a database of vector data and create maps on the screen on the fly for you to use. These "vector maps" have a number of advantages for the user. Some of them are listed below:

- They can be stored in less memory than a traditional map picture.
- They are zoomable
- You can control the amount of detail you wish to view.
- They can be searched for certain data via text searches
- They can be searched graphically such as when you click on an icon or other object on the screen.
- Text can be moved automatically as needed to remain on the screen during pan and zoom.
- They integrate well with the internal waypoint, route, and track data, which is stored in a similar fashion.

Generally speaking, the text data stored in Garmin database can be reached easily via its graphic representation on the map page. For example information on a point of interest at a particular exit on a freeway can be found by selecting the exit symbol on the map. Some data may also be searched from the menu system. Except for the Jeppesen database (for use on aircraft) the Garmin databases are only available from Garmin or a Garmin dealer. There are no third party sources for these maps or other data.

181

City Databases

Many of the older 12 channel non-mapping units support a city database. These city databases are really just a collection of items, similar to standard waypoints, that mark the location of cities. These can be useful for high level orientation when flying, or traveling on the highway. It is not clear exactly where in the city the waypoint depicts but it is likely the geographic center; as it clearly does not mark any prominent place. For non-mapping units there are 6 databases available and are selected when you buy the unit. They are not changeable or upgradeable. All of the databases include major cities (greater than 200,000 population) for the whole world and smaller cities, towns, and villages for the region of your choice. There are typically more that 22,000 locations included, sometimes a lot more (30,000 on the 12CX). The six regions are North America, South America, Europe, Africa, Asia, and South Pacific.

The city names are not subject to the standard 6 or 10 character limit of Garmin handhelds. They are shown on the map screen like user defined waypoints but cannot be used directly for navigation. If you want a city in a city database to be used as a goto destination or as part of a route it must be copied into the user waypoint memory. This is most easily done by displaying the city on the map page and then highlighting it using the pan option. Once highlighted you can hit the mark key or the goto followed by enter to create a user waypoint with the same data. In the case of goto you will also start navigating toward that waypoint. Since the city names are likely to be larger than 6 characters the GPS automatically creates a short abbreviated name for the city.

To locate a particular city in the database you would use the "Find City" command from the main menu. Once selected you should enter the desired name as if you were entering a name for a waypoint. The first entry in the database that matches the name you entered will be displayed along with the state. If there are duplicates you can use the arrow keys to scroll until you reach the one you want. Alternately you can visually locate the city using the map screen and highlight it and press enter to select. Once selected you can check the distance to the city or between cities and a reference waypoint using the 'ref' section of the page. All underscores represents your current position or you can select any waypoint in your list. Once satisfied, press enter to display the great circle distance and a bearing to the city. If a city is displayed you can also press goto to automatically create a user waypoint at that location and begin navigating toward that location. The 'showmap' entry at the bottom of the screen can be used to display the city location on the map page.

Map screen display options for the city database (city setup) can be reached on the map page local menu (called Opt on some units). You can control exactly when cities will be displayed by setting the zoom level based on the city size. When you zoom out further than the set limit that size city will no longer be shown. The largest cities in the database are always displayed. In addition another copy of the find city command is on the map screen local menu.

Base Maps

Receivers that are capable of viewing maps always include a basemap with the unit. This map provides main road coverage and a collection of city locations similar to those described above. There are several basemaps available on Garmin units. The original mapping receiver was the G-III. It came in two versions, an Americas version covering the North and South American continents and the International version that covered the rest of the world. These map choices were made at the time of purchase and were specific to where you bought your unit. The G-III pilot also had these choices for a base map. Later, when the G-III+ was released there was a new basemap that also included freeway exit information and is considerably newer than the original maps. This map is available on the III+, the 12Map, NavTalk, the Street Pilot Series, the emap, the GPS V, the 76Map, the Legend, the Vista, and some other mapping receivers. Originally this was only available in the Americas version but an Atlantic basemap was soon added for all of the above units. It provides more detailed coverage of Europe. Most recently a Pacific basemap was released which adds the data for most of the rest of the world. Like the original released maps, the basemap that is in the unit you purchase is dependent upon where the unit is purchased.

All of the basemaps include Country outlines and major city locations for the entire world. In addition they include more detailed information about the specific region covered by the unit you purchased. This information is stored in a rom and is not changeable or upgradeable. At one point the Americas data in the original III was revised so there are two versions of this unit as indicated by the map revision data on the opening screen. The newest of the GPS III version maps is also used in the Americas Street Pilot but is out of date compared to the GPS III Plus base maps.

All of the USA coverage in the base maps is based on the census bureau tiger database. In the last year or two there have been many improvements in this database that are not reflected in the basemaps. Garmin has not disclosed the version of the tiger database that was used when making these maps. The original basemaps seem to be the level of the 1995 tiger database while the III+ seems to have the 1997 update. Coverage for the rest of the world comes from various sources with varying degrees of completeness and accuracy. You can expect some missing roads and some roads inaccurately represented in these databases.

Garmin specifies the coverage of the basemaps for the Americas to include:

- Oceans, rivers, and lakes (greater than 30 sq. miles) greater than 5 sq. miles in selected areas such as the USA and Canada. Major streams in the USA.
- Principal cities and many smaller cities and towns. Principal urban areas in the USA.
- Major interstates and principal highways. Some local and state roads in the USA particularly in urban areas.
- Political boundaries (state and international borders)
- Railroads in the USA and Canada.

The international version contains similar data. The Atlantic database contains detail for Western Europe that is similar to the additional detail shown above for the US and Canada. Similarly the Pacific database provides some increased detail for Australia and Eastern Pacific rim countries.

When Garmin decided to release half the world in a GPS unit there had to be compromises. There is only so much room in one ROM so they chose several ways to reduce the size of the database.

First they decreased the detail. This is the most obvious method. Secondly they reduced the faithfulness to curves and jogs in the road. This reduces the number of points used to describe a road and thus reduces the total size of the database. A road description consists of these points connected by lines, hence the term vector database. Finally they reduced the number of bits used to describe the points themselves. Since data about the points is a floating-point number you can simply reduce resolution by a bit or two and save significant amount of space in memory. Garmin takes advantage of this by reducing the precision of the road data, and they reduce the precision even more on the background lakes and shoreline data.

Since the data has been reduced in precision you should not depend on the locations for absolute accuracy. Garmin indicates this by displaying the word 'overzoom' just below the scale indicator on the map page. The word overzoom appears anytime you have set the map resolution beyond the accuracy of the map itself. You may still wish to do this since the accuracy of the GPS itself is higher, but realize that the position relative to the map objects should be suspect. The overzoom indication was confusing to some folks so Garmin added an error circle. Basically this circle factors in your position accuracy (as indicated by your epe) and the map accuracy to display a circle of uncertainty. You are probably somewhere inside the circle as displayed on the map. If you turn the map display off (on units with this capability) then the circle will get much smaller as it will display the precision of the fix itself. By the way, this is just an estimate, you might still be outside the circle. The circle may be turned on or off as desired except on the emap where it is always on.

The basemap contains the equivalent of the City Database for mapping receivers. It can be searched and used in a similar way to the discussion above for non-mapping receiver. Units with a find key offer a city search feature when you press this key. Note that there is a world wide map that includes the name of some major cities but the search function will only find cities in the detailed area of the basemap.

Loadable Maps and POI

Garmin provides a Points of Interest, database on all mapping units and a few of the newer non-mapping units. This section covers both kinds of units. For example, there are exit POI's in the basemap for loadable map units. An example of an exit POI is shown below (exit 225).

Figure 22 Exit POI

Exit POI's can be reached by highlighting the small square showing the exit and hitting enter. Then use the keypad to select the item of interest and press enter to display the POI's. These POI's includes many businesses within about 1/4 mile of the exit, including; restaurants, diesel/gas, hotels/lodging, overnight RV parking and dump, campgrounds, truckstops, medical facilities, shopping and outlet malls, ATM's, and many more attractions.

Loadable Maps are supplied on a CDROM called a MapSource CD and are loadable via a program that is included on the CD. The program on the CD is often out of date so you should visit the Garmin web site and download the latest version. This program can be used with mapping units and non-mapping units to upload and download tracks, routes, and waypoints however map databases can only be loaded into mapping capable units. In addition to the maps themselves the cdrom contains POI information to varying degrees based on which cdrom you purchase. Note that only MapSource CD's can be used to upload POI and mapping information to Garmin units. Use of the MapSource product is covered in a separate chapter.

USA

There are basically three completely different groups of MapSource cdrom data for road use. These include City Select, MetroGuide, and Roads and Recreation. In addition there are some more specialized mapping products available such as Topo maps, Waterways and Lights for marine use, Fishing Hotspots for specific lake coverage, and Blue Charts for serious marine use.

The MetroGuide cdroms began originally as preloaded cartridge maps in the USA and are designed specifically for GPS upload into handheld devices. These MetroGuide maps in are road and street level maps using the database from ETAK Corporation (a subsidiary of TeleAtlas). They are very accurate and contain a significant amount of POI data, primarily of commercial establishments, and street level addressing for most areas. In addition Garmin has designed specific features, such as road lock (where the GPS position display locks to the road on the map) and POI search facilities. These maps cover the full USA. However, not all of the data has been verified by ETAK. Some 20% of it is the same as the Roads and Recreation data. Originally they were packaged in sections that were too large for most hand held units except the emap. A newer version has been released that has improved coverage and smaller segments permitting these to be used in all Garmin mapping receivers. The newer version also includes routing information in the database and can be used with the Garmin V to provide directing automatic routing on the unit and with MapSource to provide automatic routing on the pc. In addition to mapping visual support these map databases provide street level address searching and next intersection information. They also provide the support necessary to have a unit lock onto a road and track your position with respect to that road.

The City Select maps are shipped with the GPS V product and are only usable on this handheld and the larger units. These maps are provided by NavTech, which is also the supplier of most of the maps that are used in the dedicated vehicle based systems. The original USA release only covered major cities and interconnect roads, but the newest release covers the full USA with street level detail. It provides all of the features of the MetroGuide line of products and generally has more information for the autorouter. The autorouting functions in some of the outlying areas leaves a bit to be desired since the release seems to have be rushed out by Navtech.

Another USA road map product is the R&R, Roads and Recreation, cdrom. It contains detailed street level data and recreation POI's. It also contains updates to the basemap freeway exit information. The street level data comes from the Tiger database which is the same source as is used on the base maps. However, in the case of the R&R cdrom it is the full database at full resolution. Note however that the Tiger database itself is not maintained to the same degree of accuracy as the ETAK or NavTech databases of the same area, and Garmin has never updated their original product. Due to when it was released it has data in it that is even older than the base map data in some of the newest units, probably based on the 1995 Tiger data. The resolution of this data can be observed by turning on the error circle. You will find the circle significantly smaller than the base map circle but, for USA data, it is larger than some of the other databases available from Garmin. The map downloads are usually divided by counties and you can generally load several maps into the 1.4 Meg space of the older Garmin units.

Other US map cdroms include a topographic cdrom and most recently a marine database. The topographic cdrom is derived from USGS 100K data whch is very precise, equaling the ETAK data, but can be rather old since USGS surveys are not done very often. In some case the data can be as old as 20 years or more. The topographic data for government maps are often updated via aerial photographs so will often not include updates to man made objects. The topographic data includes elevation data and roads, but roads that are generally not named, and includes some trails. It is most useful for off-road and backpacking use. The maps contain the highest precision of any of the MapSource maps available for handhelds and rival the precision of MetroGuide maps. The topo maps include everything that is on the R&R cdrom for the US except road names and freeway exit data. They include many items not found in the R&R database such as good coverage of coastal navaids (sorry not inland), pipelines, airport runways, towers, and other physical features. They even include some underwater contours.

There are also marine products available. The Waterways and Lights cdrom for the US includes improved coverage of outlines and navaid data for coastal and inland areas. It focuses on shoreline information while providing some road coverage. A future release is supposed to include some depth contours. See the marine navaid section below for more details on navaids. The most complete charts for Garmin receivers are on the Blue Chart cdrom. This is a very accurate but expensive set of nautical charts. For inland users there is a Fishing Hotspots cdrom that features data on many of the inland lakes.

International

The maps for Europe are all from NavTech and all provide the same level of road detail but differ in other respects. The City Select maps provide autorouting data on the GPS V and on the PC while MetroGuide maps only provide autorouting on the PC. Finally the Roads and Recreation maps provide no autorouting but are the smallest files sizes. They do not support the street address database lookup. Originally MetroGuide and Roads and Recreation maps were sold on a country by country basis but the newer release has the entire continent in one set. City Select requires an unlock code for the various regions. One is supplied with the product but an additional unlock must be purchased for the rest of Europe.

Canada has maps similar to those of Europe and are also from NavTech. Only major cities are covered but both the MetroGuide and Road & Recreation are on the same cdrom so the user gets both sets. In addition there is an enhanced basemap that covers areas not covered in the NavTech maps.

Australia has a set of maps as well, also from NavTech. They include City Select and MetroGuide. There is a new set of City Select maps for South Africa.

For areas outside the ones identified above, a users only choice is to buy a cdrom containing the Worldmap. This is basically a collection of the various basemaps so that a user can have maps in the area of the world not covered in their particular basemap. This database is pretty old and contains about the same resolution of information that was found in the old G-III international version basemap for the whole world. It provides Americas basemap level coverage including freeway exit data, which is useful for international customers. It does provide coverage of road level data in areas not covered by any other MapSource maps to date. It has not been updated even to the level on the newer G-III+ Atlantic version. It is probably better than no maps at all, but not by much. Hopefully Garmin will update this with the newer data that is now available. It does have the best source of worldwide coastal marine navaids and good coastal maps.

General

Garmin has made a recent commitment to update the Mapsource maps yearly and are beginning to deliver on this promise although not all of the maps are updated anywhere near this often. You can view the maps on a per cdrom basis at the Garmin web site to determine if its contents are suitable for your area.

The GPS receivers that support maps also support the ability to upload multiple maps simultaneously up to the limit of available memory. You can even upload maps from different MapSource cdroms. Each time you upload you must load all of the maps you want as you cannot add maps to an existing upload. If a map covers the same area as a basemap then it will automatically be used in lieu of the basemap when the GPS determines that you are located in an area covered by the uploaded map. If there are two uploaded maps covering the same area then the results always follow a prescribed priority. City Select is the highest priority, followed by MetroGuide, etc. down to World Map. The receivers provide the ability to turn off the display of individual uploaded maps so that you can see the map you wish. Some receivers can also turn off a whole category of maps such as all Topo Maps. Control of maps is done with the local menu on the map page. Select the MapSource info tab to display information on loaded maps and to select which ones should be displayed. The area and detail, including poi's, in an uploaded map is predetermined by software and is not adjustable by the user however, once the map is loaded, you can adjust the display of this data.

You can search for cities on the maps using the "spell and find" command. It is located on the main menu under cities. You use it just like the find cities command described above in the city database section. If you wish to use a city as a destination in a route you must convert it to a standard waypoint as described in that section. There is an option of the spell and find local menu that can be used to display the city size or use it as a reference to measure distance. To view any services or POI's you would select them from the map page.

The emap and newer units have a more sophisticated search capability using a dedicated find key. Pressing find will bring up a menu that permits you to find waypoints, cities, or freeway exits. If you have a MapSource database loaded you will also be able to search for poi data. If you have an optional MetroGuide map or city select map installed you can also find more points of interest, intersections and addresses. Select the menu item you wish to bring up the selection screen. Selection screens can have a local menu that permits switching between find by name and find nearest. Use the arrow keys to select the one you want or to enter the name of the one you want and press enter to find more information about the item. Exit data (and POI data for installed maps) has more menus to permit selection of an individual point of interest. Once the poi is selected you can press enter to find information about it including directions to find it from the freeway exit or its address. Once an item is selected it can be the target of a goto by selecting this entry from the screen. Unlike other handheld units the found items do not have to be converted to waypoints to use them for navigation. You can, however, use the local menu to convert them to waypoints if you wish.

Generally map objects will be found on the currently selected map. In some cases you might want to search a different map that the one you have selected. For example, since the basemap and the downloaded maps both contain city names there needs to be a method to determine which map to search. There are two different scenarios that are used for search. You can search by nearest and search by name. When searching by nearest the default is to search using the currently selected map. You should try the local menu on your unit to see if a choice of maps is given. When searching by name the default is always to search the basemap. Some versions of software and products will only search the basemap, or the worldmap when outside the basemap region. But newer products may let you select the map you wish to search. When you search by name the name entry for the selection is automatically selected. OK or cancel this selection to be able to get to the local menu and then check to see if there is a map selection option.

Jeppesen

The Jeppesen database is a service that contains a worldwide database of information needed by pilots. A Jeppesen database is supplied with the III Pilot and can be updated either with one-time upgrades or through a subscription service. These can be obtained from Garmin or directly through Jeppesen. The current effective date for the database in the unit is shown on opening database screen.

The Jeppesen database contains information on airports, runways, communications frequencies, VORs, NDBs, intersections and airspace boundaries. This information is available on the map page and from the waypoint menu choice on the main menu. In addition certain features of the III Pilot can take advantage of the data in the Jeppesen database. For example, the vertical navigation feature can use the airport location and altitude data. (See the navigation chapter for a discussion of vertical navigation.)

To access the database information from the map screen, use the rocker keypad to place the panning cursor over the icon containing the desired information. The identifier will be highlighted. Press enter to view the information. If an airport is selected you will have a set of tabs that can be used to select airport, runway, or communications pages. If the cursor is in an open area of the map you will get airspace information if the enter key is pressed.

To access the database from the main menu hit menu twice as normal, highlight 'Waypoints' and press enter. A series of tabs will appear that lets you select from airports, runways, comm frequencies, VORs, NDBs, intersections, or user-entered waypoints. Select the category with the rocker keypad. The information can be searched by identifier or, in the case of airports, VORs, and NDBs, by facility name or city. Highlight the appropriate field, press enter and supply any needed name by using the rocker keypad.

The map page contains icons representing the Jeppesen data and also shows the airspace boundaries graphically. This information can be turned on or off using the map display menu. The available settings include Aviation DAta on the map menu, large/medium/small airports on the Apt menu, VORs/NDBs/intersections on the Nav menu, Controlled airspace on the CTRL tab, and Special Use airspace on the

SUA tab. The highway display also shows the airports that may appear along the route you are traveling.

The 9 nearest airports can be located using the GOTO/NRST key. Press and hold the key until the nearest page appears and then select the airports tab. You can find information about the airport by selecting the entry to view it. You can also highlight the desired airport and press the GOTO key followed by ENTER to designate this airport as your desired destination.

Specific information in the Jeppesen database includes:

- Airport - Identifier, city/state, name, position, elevation, fuel
- Runway - Length, width, orientation, surface, lighting, diagram
- Comm - Frequencies for: ATIS, pre-taxi, clearance, ground, tower, unicom, multicom, approach, departure, arrival, class B, class C, TMA, CTA, TRSA
- VOR - Identifier, city/state, facility name, position, frequency indication of co-located DME or TACAN
- NDB - Identifier, city/state, facility name, position, frequency
- Int - Intersection name, region/country, nearest VOR
- Station - FSS or ARTCC points of communication, up to 5 for nearest stations.

Data that can be displayed graphically such as locations and airspace boundaries will be shown on the map page except runways have a special graphic display as part of runway information.

Jeppesen data can also drive the alarms so that you are warned when you change from one control space to another or entry into a restricted area, proximity to control towers, etc.

If there are conflicts between the Jeppesen names and user defined waypoint names the III pilot will bring of a list of duplicate names so that you can choose which one you want.

Marine Navaids

Navaids are the aids to navigation that are used in marine applications. (Air navaids are included in the Jeppesen data shown above.) These include such things as navigation buoys, beacons, fog horns, and light houses. Of the handheld units covered in this manual only the Garmin 48 includes navaids as part of the base product. Most of the information in this section is based on this unit. However, the units that support loadable POI information also support loadable navigation aid information. The Garmin 48 is updateable using the Waterways and Lights cdrom. The 48 is divided by region exactly the same way as the city databases described above. On mapping units the navaid data is treated similarly to the POI information available on the 'Roads and Recreation' CD's and the 'WaterWays and Lights' CD.

Some products such as the etrex venture, legend, vista and 76 line come with a USA Navaid database pre-installed in the poi or map memory. This database will be overwritten the first time the user uploads any maps or poi data. Garmin has a copy of this database in a self-installing file on their web site that can reload the navaid database directly to the unit. Of course, this would overwrite any poi's or maps the user loaded. The poi database upload program and many of the mapping products include navaid information so this is generally not a problem.

The display of navaid data on the map page is under control of the local menu navaid setup. You can control exactly when objects appear based on the zoom level. Control is based on the range of the object. Some Garmin units do not distinguish navaids from other poi's and may not have this feature. Navaid labels are also configurable from navaid setup. Unlike the city database navaids can be treated like a user waypoint on the 48 but must be translated to user waypoints on other units to use them for navigation or alarms.

Navaid icons on the G-48 are shown below.

Foghorn	
Racon	
Radio Beacon	
Daybeacon (red triangle)	
Daybeacon (green square)	
Daybeacon (white diamond)	
Lit Navaid (color indicated in symbol)	
Lit Navaid (multicolor)	
Unlit Navaid (color indicated in symbol)	
Unlit Navaid (multicolor)	

Figure 23 NavAids

The label on the navaid is coded to supply information about the navaid itself. The following is the decode of the label data:

Lit Navaid Abbreviations
   F - fixed - continuous and steady light
   FL - single flashing
   FL(2) - Group Flashing - group of repeated flashes
   FL(2+1) - composite Group - complex light
   FFL - fixed with brighter flashing light
   Q - quick flashing 60 flashes per minute.
   VQ - very quick flashing
   OC - occulating - longer light than dark
   ISO - Isophase - light and dark equal

MO(A) - Morse Code (letter to be sent)
AL - Alternating different colors
DIR - Directional
Unlit Navaids
S - square
J - Junction
T - Triangle
K - range
M - safe water
C - crossing
N - special purpose
Colors
G=Green - R=Red - W=White - Y=Yellow - B=Black
Vi=Violet - Or=Orange - Bu=blue - Am=Amber

The new mapsource "WaterWays and Lights" cdrom, the "topo" cdrom, the older "Roads and Recreation" cdrom and the "Worldmap" cdrom contain navaids as well. These include some wrecks and other potential obstruction data not included on the G-48. They can only upload navaids as part of their generally mapping uploads for mapping receivers. Coverage for the W&L cdrom includes inland bodies of water as well as coastal areas. It provides coastal and lake outlines and navaid data. Other cdroms have differing coverage depending on the map cdrom. Check the map section above for details on coverage. The types of navaids available on the cdroms include all of the types shown above for the G-48 plus ship wreck, rocks and other hazards to navigation. The figure below shows the icons and the information they represent.

LIGHTED NAVAID, MULTI-COLORED   DAYBEACON, WHITE DIAMOND

LIGHTED NAVAID, BLUE   DAYBEACON, GREEN SQUARE

LIGHTED NAVAID, VIOLET   DAYBEACON, RED TRIANGLE

LIGHTED NAVAID, ORANGE   RACON OR RADIO BEACON

LIGHTED NAVAID, YELLOW OR AMBER   FOG HORN

LIGHTED NAVAID, GREEN   WRECK, CLEARED BY WIRE-DRAG

LIGHTED NAVAID, RED   OBSTRUCTION, CLEARED BY WIRE-DRAG

LIGHTED NAVAID, WHITE   SOUNDING

UNLIT NAVAID, MULTI-COLORED   ROCK SUBMERGED AT LOW WATER

UNLIT NAVAID, ORANGE   ROCK AWASH

UNLIT NAVAID, YELLOW OR AMBER   OBSTRUCTION, SUBMERGED

UNLIT NAVAID, BLACK   OBSTRUCTION, AWASH

UNLIT NAVAID, GREEN   OBSTRUCTION, VISIBLE AT HIGH WATER

UNLIT NAVAID, RED   SUBMERGED WRECK, NON-DANGEROUS

UNLIT NAVAID, WHITE   SUBMERGED WRECK, DANGEROUS

UNLIT NAVAID, OTHER OR UNKNOWN COLOR   VISIBLE WRECK

NAVAID, UNKNOWN CLASSIFICATION   UNCLASSIFIED OBSTRUCTION

Figure 24 More NavAids

Chapter 13
Working with MapSource

Introduction

This discussion is based on MapSource version 4.10. MapSource is released quite often with new features so this information may not be current.

Garmin requires that users use their product called MapSource to upload maps and poi's (points of interest) to their GPS units. As a matter of fact it accompanies the GPS V unit when it is purchased. This chapter was written to cover the use of MapSource with handheld GPS receivers since it is shipped with this model. This is not an endorsement of MapSource as the tool of choice for waypoint maintenance but it does that job as well. In a general way this can be used as a guide for other third party tools. This chapter is not intended to replace the Garmin documentation but will provide some tips that may help in understanding the most effective ways to use this product.

MapSource is the name of the software program itself but this name is often used interchangeably with the names of the various maps. This can lead to confusion particularly when dealing with features differences that are map specific. This chapter will always use the term MapSource to refer to the program. This program can provide three separate but related functions for the GPS user. It can upload purchased maps and/or to the GPS device, it can manage user data such as waypoints, tracks, and routes, and finally it can provide a real time tracking ability when the GPS is attached to a pc while you are traveling down the road. In addition the program has the ability to aid in route planning depending on which maps are owned and being used by the user. MapSource can provide autorouting as well if the underlying database has the appropriate data. The resultant route can then be downloaded to the GPS even though it does not have autorouting capability.

Version 4.10 of MapSource adds the ability to support languages other than English. Languages supported include French, German,

Italian, Spanish, Japanese, Korean, Chinese (Traditional), and Chinese (Simplified). This support includes international map products.

Installation

Product installation on a pc is pretty simple using the standard setup command. If you try and load a second product it will sense that an existing product is loaded and install only the map information. Mac users have indicated that a pc simulator can be used with this product although there have been mixed results with the actual download of the map data to the GPS. Generally you may have to run the serial interface at less that full speed on mac simulators.

If you have the latest version of MapSource on the cdrom you will be given a choice of a full install. Selecting this will cause the entire cdrom to be loaded on your hard drive such that the cdrom will not be required to run the program. In addition the city select and newer MetroGuide product also installs all or most of the maps so that autorouting can be performed on the cdrom. However, most product installs require the cdrom be used to contain the data. If you would prefer to be able to run without the cdrom on one of these older versions of MapSource simply copy the entire cdrom to the hard drive and then perform the setup directly from the hard drive copy.

Once Map Source is installed, use the about button under the help menu to check the version. Garmin releases new versions of MapSource frequently, which can be downloaded from their web site. However, the cdroms are not re-released when the program changes so it is very likely that the cdrom installed a down level version. The download from Garmin requires a previous installation of MapSource to install. The about button is also the place to check which map products are loaded and the version of these products as well. As with the GPS units Garmin always releases a product as version 2.0. Versions 1.X are used for testing.

A few of the map sets sold by Garmin use unlock codes to enable the maps. These unlock codes are keyed to the GPS itself and will support up to two GPS devices for each key. If the map set you purchased requires an unlock code the installation will prompt you to run the unlock wizard. You can run this during installation or later.

Generally unlock codes can unlock a region of the cdrom or the full cdrom depending on what you purchased. Often a cdrom purchase will include only one region in the base price with additional regions requiring additional purchases.

Overview

When you start MapSource you may need to have the cdrom already loaded so that MapSource can use the maps that are on it, however the equivalent of the basemap is already present on the hard disk so some map viewing can be done even if the cdrom isn't loaded. Look at the view menu and find the entry for 'show GPS map detail'. This is a toggle and, if set to off, you can view the basemap and any zoom level without needing the cdrom. The screen will look something like this:

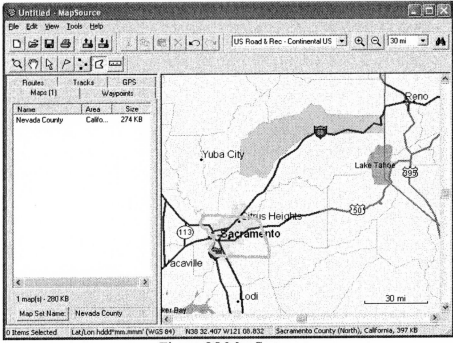

Figure 25 MapSource

201

The standard menus of file, edit, view, tools, and help are provided. Many of the frequent commands from the file and edit menus are available as icons on the next row followed by some favorite viewing options. Much of the time you will be using the tools menu which is duplicated as icons on the third row. The left side contains a tab separated text area while the right portion shows the map itself. There is a movable bar between these two items. A status bar appears at the bottom of the screen.

MapSource is only able to work with one database at a time but it is possible to run multiple copies of MapSource, which can use different databases. It is possible to cut and paste data between these separate copies.

Working with Maps

The primary purpose of Mapsource is to select and download maps to the GPS. In early versions of Mapsource this is about all the product would do. And it still is the only thing that you can't use another program to do.

The maps must be loaded as a set to the GPS since a full table of maps is created by MapSource. There is no provision to update a map on the GPS incrementally and thus there is no need to erase the map memory before starting the transfer. You can erase the map memory if you wish by downloading a map set that contains no maps. Some of the GPS units include a memory capacity bar but this is not used for map memory although it is used to store the table of map contents.

Download Step

To use the product to download maps perform the following steps:
1. Start the MapSource program
2. If necessary, Place the cdrom containing the desired maps in the cdrom drive and select the correct region from the view menu or the pull down choice on the second row.
3. Use the pan tool (or the scroll bars on the map edge) and the zoom tool (or the view menu zoom items) to get the desired area on the screen.
4. Use the tools menu or tool bar to select the map tool.

5. When you move the cursor over the map using the mouse the map area choices will be highlighted in yellow as shown in the figure above for the county of Sacramento. The status bar will show the name of the selected area.

6. Use the left mouse button to select the map you wish. This will cause it to change colors and it will show up on the left area list as a selected map.

7. If you can't see the map in the list click the map tab above the list area to bring this into view. There are three columns, the map name, the area the map is in, and the size. You can use the mouse to move the column widths or center bar so that they are all visible.

8. Repeat the map selection steps as needed to select the maps you wish. Note that clicking a previously selected map will remove it from the list.

9. The order that you click the maps is the order that they will appear in this list and in the list that is in your GPS so you may wish to be careful to select them in the order you want. You cannot move them around once they are on the list, however you can sort them if you wish. They can be sorted by name, location, or size by clicking the banner for the sort you wish. Clicking again reverses the sort order.

10. At the bottom of the list is a total size for the maps you have chosen. Watch this number to ensure that you do not exceed the memory capacity of the GPS device you intend to download to. Common capacities are: 1.4 Meg on the III+, 8 Meg on the Legend and the Map76, 24 Meg on the Vista and 76S, and 19 Meg on the GPS V. The emap capacity depends on the cartridge size you purchased.

11. You can mix maps from different cdroms by just changing the cdrom and then selecting the one you wish from the pull down. Later during the actual download you will be prompted, as needed to install the correct cdrom.

12. When you have selected all the desired maps you are ready to perform the download to the GPS unit.

13. Name the map set as desired by clicking on the map set name button at the bottom of the list screen or use the default name of first and last map.

14. You may wish to save this list of maps on your hard disk for future use. Use the Save icon or the File menu Save As command to bring up the save form. Name the file as desired and click ok. This will create a .mps file on your disk.

15. Click the Save to Device icon or the File menu item to bring up the save to Device Form.

16. Use the Serial tab if you intend to download to a real GPS even if it is hooked to the computer via a serial to USB adapter. Use the USB tab only for downloading to the Garmin USB cartridge tool.

17. Make sure the GPS is hooked up and is set to Garmin or Garmin/Garmin mode from the main menu interface choice. It is best to set the GPS to simulator, use indoors, or GPS off mode to avoid any conflicts with it trying to do two things at once. If you are running the GPS from batteries, make sure you have a good set installed before trying to download maps.

18. Click auto-Detect to test the interface connection and resolve any problems before proceeding. Generally the fastest baud rate is the one you want but if you have problems with the download you may want to retry by selecting a lower baud rate.

19. Select the button turn off after the download to save battery power. Do not select any transfers except map. Otherwise you may mess up your routes, waypoints, and tracklogs. It is possible to set up a mps file to transfer everything at the same time but it is recommended that you keep it simple for now. Note that routes, waypoints, and tracklogs are always transferred at 9600 baud no matter what baud rate you select.

20. Select Save to perform the transfer. This will take a long time if you have several maps to load. Be patient.

21. Resolve any error messages you see. If the memory capacity is exceeded you may need to remove a map or two and retry. If you get transfer errors you may need to lower the baud rate and retry.

As stated in the above steps you can mix maps from multiple map sets in the same download. You cannot alter the maps or poi information. The individual map size is fixed. Garmin does sell a special poi database product that allows you to custom select poi's and this can be mixed with other map sets for a download. Some products support poi's but not maps. These products will not display maps even if you manage to download some to them.

There is one preference setting that affects map download. If you are not using a GPS V but have maps sets that would expect to download autorouting data such as the Metroguide USA maps, you can significantly reduce the download size of the map data by not downloading the autoroute data. To do this, select the preferences menu item from the edit menu and select the transfer tab. There is a check box to choose to delete this data from the download. Other preference items include the default com port and default baud rate.

Unlocking maps

Some map sets require unlock codes in order to use the maps with a GPS. These codes are keyed to individual GPS units Garmin permits one purchased unlock code to be used with up to 2 devices. Most map sets do not require unlock codes. If you need to unlock a map you will also need to determine the GPS id code. Both the unlock form and the id form are on the tools menu. There is also an unlock wizard installed if you install a product that requires unlock codes.

If you receive the long 25 character unlock code you can enter it directly into the unlock form. This form is also the place to display map sets that were previously unlocked. You may have received an 8 digit coupon instead of the full unlock code. In this case you will need the unlock wizard which can be launched from windows on the start menu by selecting programs and finding the MapSource entry. The unlock wizard can use an Internet connection to download the unlock code.

To obtain the id from the Garmin receiver you will need the receiver set to Garmin or Garmin/Garmin mode.

The unlock form can also be used to export or import an unlock database. It is a good idea to backup your unlock database onto a floppy or other safe media.

Working with Tracks, Waypoints, and Routes.

The MapSource product provides the ability to manage the Track, waypoints, and route databases that are inside your GPS. This can provide for backup of this data and will also provide for the ability to edit this using a keyboard and mouse which is usually much easier than trying to do it on the GPS itself. There are many other programs that can also be used to manage this data but Mapsource provides some unique features that allow it to directly work with the same maps that you can download to the unit.

The ability to work with tracks, waypoints, and routes is available for all Garmin units, even if they do not support maps or poi's. However to get the product itself the use must purchase a product containing maps or poi databases and the map on the pc can still be used for real time tracking or to help in route and waypoint creation.

Prior to releasing MapSource Garmin had another product called PCX5 (Dos based) that could be used to manage tracks, waypoints, and routes. This product has been discontinued but there is still the ability to import databases created with this product to Mapsource. The import menu selection on the file menu offers this option. Unfortunately you cannot export files in this format, nor can you import files in any format that you can export so MapSource does not permit working with its data outside of the MapSource environment. As a work around some folks actually download MapSource data to their GPS and then upload it into another program or vice versa to do data exchange.

Waypoints

Waypoints can be downloaded from the GPS and modified in MapSource or they can be created directly in MapSource. The easiest way is to create them graphically on the map itself. If you are using the waypoint tool you can just click to place a new waypoint. You can also right click and select new waypoint at any time to create a new waypoint. Once the waypoint is selected a form will appear that allows you to modify any of the data for the new waypoint. Click OK to create the point or cancel if you decide you don't want to create a new waypoint. There are several fields on the form that may or may not transfer when the waypoint is sent directly to the GPS. For example some GPS units do not support altitude while others do not support the description data; the oldest units don't support icons. If the GPS doesn't support some particular field it will simply be ignored during the GPS upload.

There are several preferences that effect waypoints and waypoint generation. You can select the grid and datum you wish to use to display the location and the units for altitude and depth. You can also pick the default prefix to be used in name generation as well as the length of the name. Note that older Garmin units only support a 6 character name while the newer units support a 10 character name. The units, position, and datum information are only for the purpose of display and data entry. They will be converted to an internal format when uploaded to the GPS so it really doesn't matter which you use. Leave the datum at WGS84 unless you are trying to match information from a paper map that you are using as a reference.

Routes

A route represents a collection of waypoints/mappoints connected in a prescribed order. It defines the desired path from one place to another. Generally it identifies the critical turns that need to be made by the user to get from their starting location to a destination and the GPS can be used to provide directional guidance from one point on the route to another. Routes are usually created manually by the user as part of the planning process for a trip, but they can also be created automatically by Mapsource if the correct mapping database is used or they might be created on the GPS itself and stored for later use by Mapsource.

The unique thing about routes made in MapSource is the fact that they can take advantage of mappoints on some units. Mappoints are locations on mapping receivers that are identified and reference to map locations and are not waypoints. Thus they do not appear in the waypoint memory and are not limited to the 500 or so locations supported as waypoints. (They do effect the amount of memory reported in on the memory used display.) The user will need to be careful here since some Garmin units do not support mappoints and for these units the mappoints will need to be converted to actual waypoints. This can be done by right clicking on a mappoint and selecting waypoint to create a waypoint from the mappoint data. The MapSource documentation does not always make the distinction between mappoints and waypoints clear.

Generally a route is built in MapSource by clicking on successive map locations and these are mappoints by default so long as the point clicked contains a mapping object. Once the desired route is created it can be uploaded to the GPS and used for navigation. When building a route the user will need to consider the capabilities of the GPS unit that they own. For example, routes have a limit to the number of total points that can be contained in one route. Many units limit this number to 30 while some units allow up to 50 points, and the GPS V can support up to 250 mappoints. If you build a route with more than the prescribed number of points it will be truncated when you load it to the GPS. In these cases you will need to split it into several routes or remove some of the intermediate points.

Another way to work with routes is to use the route properties (select with right mouse click or from edit menu) to add, delete, or edit property data. An existing route can be duplicated as a starting point for a new route or as a step in splitting the route.

It is possible to reuse the same waypoint or mappoint in more than one route and this is often done when splitting a route into sections. Routes and portions of routes can be cut and pasted inside the MapSource product or even between two copies of the program running simultaneously.

Tracklogs

Tracklogs cannot be created manually in MapSource but it is possible to do some editing on the tracklogs that are downloaded from the GPS. In addition it is possible to create a tracklog using the real time display capability (described below). All Garmin units support an active tracklog but many also support up to 10 saved logs in addition to the active log. When uploading the log back to the GPS it must be named "Active Log" in order for it to replace the active log, otherwise it will be stored in one of the saved logs on these units.

Tracklog data can be displayed in MapSource and will include some information that is not directly stored in the tracklog. Active logs include information about the time and location and can include breaks. Some logs will also include altitude data. MapSource can use this information to compute speed data, leg length, and direction, which can be shown using the track properties on the right mouse button. It is even possible to display a vertical profile for logs containing altitude data. Saved logs do not include the time stamps so the speed data cannot be calculated. MapSource will also show breaks in the active log as if there were multiple logs but there is really only one active log.

To edit the tracklog, start by displaying the properties. You can cut and paste individual tracklog entries and you can paste from one tracklog to another. Be careful you do not exceed the length capacity of the log. If you make a mistake use the edit undo command to fix it. You can also rename the tracklog or invert it.

Downloading and uploading data to the GPS

The upload/download dialog box allows you to select whether you wish to work with waypoints, routes, tracks, maps or any mix of the above. When uploading and download data the full list is used. All of the waypoints in the GPS are downloaded at once, as are the routes and tracks, if selected. If your maps are not tied to your waypoints then it is suggested that you always work with them separately such that the maps are not stored in the same database as routes, waypoints, and tracks. This can be done simply by saving the data in separate files.

If you upload a route that uses waypoints you will need to ensure that you also loaded the appropriate waypoints as well. Otherwise the route could reference existing points at the wrong location or fail to work at all.

The oldest Garmin units do not use names for routes but rather have the routes numbered. On these units MapSource will display the name comment but uploads will start with 01 and multiple routes will be uploaded sequentially erasing any routes with the same number on the unit.

Working with Real Time display.

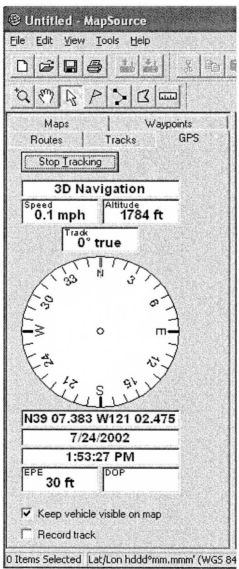

Figure 26 Real Time Display

Garmin also supports a real time display option with MapSource. Using this feature permits an attached pc to provide mapping display and tracklogging while directly attached to a GPS in real time. There is real-time display support for NMEA mode or Garmin mode. Some of the Garmin units do not support real time output in Garmin mode, but if your GPS does support Garmin mode then this is the preferred mode for real-time display. In NMEA mode the update is every two seconds while in Garmin mode it is every second which makes the display feel a bit more responsive to what is actually happening in your car. In addition, the maps that have an unlock code require Garmin mode so that they can check the lock status.

To set up the display just pick the GPS tab on top of the list display screen and then click on start tracking. The program will automatically sense whether the unit is set to NMEA mode or Garmin mode and will begin track. There are only two options. You can force the current location to remain on the screen and you can turn on tracklogging if you wish. If track logging is turned on it creates an active log on the pc in the same format as it would be if it were collected with the GPS except that it records a point at each sample of the data. If you stop the log or stop the GPS tracking then start it again it will create a new active log. The current log will be displayed on the screen as a breadcrumb track of your current movements. You will need to save the tracks in a file if you wish them to last longer than one session. If you scroll the window the keep car on screen flag will turn off. Click the flag to return the screen to the GPS location.

All of the data displayed is under control of the preference form reached from the Edit menu. You can change the display formats just like you can in the GPS itself. It will be saved automatically. Note that dop (dilution of precision) may or may not be displayed depending on whether the GPS is sending this information. You might try switching between Garmin mode and NMEA mode to see this data as NMEA generally supplies it.

While the GPS display provides real time tracking it will not display navigation data. The compass display shows your current track direction, which is the same information as the digital readout just above the display. The compass output is customizable using the preference settings.

Other Features

MapSource has many special features beyond the basic ones shown above. In many cases these features will be enabled only if you have the correct map product installed and selected. The menus and forms may have additional items and choices available to match the particular map set. For example, if the map set supports autorouting on the pc then additional menus will be available to perform this task.

Setting the PC clock

If you have your GPS hooked to the pc and tracking satellites you can use it to set the pc clock from the GPS. This provides a very accurate source of time information for the pc. Simply select the 'Set PC clock' from the tools menu.

Working with the Find command

The Find command is available as a icon (binoculars) or as a menu select item from the View menu. This mimics similar functionality that is on the GPS itself but often the GPS can find object types that will not be found using the pc product. The form that appears is dependent on the map set you have currently selected. A basic form provides the ability to search for cities in your database but an advanced form will appear if your map set supports advanced searches. Advanced search include city, poi look up, as well as street and address searches. Searches are limited to the selected map region. If you are using a map with a lock code the advanced search functions will only work in the unlocked area although you can do city searches on the full map.

Another way to find objects is just to display a map of the area of interest and move the mouse cursor over objects on the map. A popup display will appear identifying the object under the cursor.

Once you have found the desired object you can view it on the map. Even though you have selected to view the object on the map you may need to zoom the map in to actually see the object. Use the right mouse button on the object to view a menu. If the menu offers the choice of map features then select this and then select the properties choice. This will generally happen if you are zoomed out too far and MapSource isn't sure which object you are interested in. If the choice is clear to MapSource it will offer the properties choice on the first menu. For a poi this will show quite a bit of additional information about the object.

Working with the autorouter

If you own a mapset that permits autorouting on the pc you will find that Mapsource has added this feature to its menus. Once the route is created it can be treated like any other route that you might have created manually and can be downloaded into a GPS even though the GPS does not directly support autorouting. A route downloaded in this fashion is subject to the standard route limitations of the GPS itself so the names of route points could be truncated and the number of the route points in the route could be truncated during the download. You may need to split the route into several routes to load it into your unit.

Measuring Distances

MapSource can measure distances and bearings between points on the map. This is available as a tool icon or by using the right mouse button to select a starting point. Each click will add a leg to the measured distance. The right mouse button can be used to cancel the current distance measurement or to clear the measurement and begin again. The total distance and the bearing from the beginning point will be shown on the status bar.

Chapter 14
Secret Startup Commands

There are several undocumented commands available, primarily at startup, on Garmin handheld receivers. This chapter will document some of the features available in a universal way although details may differ depending on the unit you have. There are variations among models and among software releases within models. There is no attempt to provide 100% mapping to a particular model or release. This is a "try it" approach where what you may see is documented and you will need to verify exactly what you did see. You may wish to check mark the features that are applicable to your unit.

Note that Garmin does not document these features for a reason. They are generally for testing the device and are not intended for end users. The modes are not supported by Garmin and may cause you to lose data in your machines. **Try this at your own risk!**

One major feature that is revealed when using these undocumented test modes is that there is a thermometer inside your Garmin unit. The intent of this thermometer is to compensate the internal time-of-day clock for changes in temperature that will cause the internal crystal to drift. To a lesser extent it is also used to adjust the contrast of the display screen for temperature changes. It will not measure ambient temperature except when the unit is first turned on since the internal temperature is changed due to the heating effects of the electronics in the unit. The thermometer based correction works by building a table of correction values that are then applied to the internal clock. This is only used to obtain an initial fix since after the fix the satellite data is used to keep the clock accurate. Similarly while you have a fix the accurate clock can be used to update the temperature data in the table which means the unit will compensate for aging parts and even the complete loss of the table.

Special key startup sequences

The three keyboard keys on the right side of the unit have special significance if held down while powering on the unit. These keys are generally called page, mark, and enter. On the GIII units the center key is called menu but for our purposes it behaves the same. On Street Pilots the equivalent keyboard names will have to be used. These are page, option, and enter. See below for etrex and emap startup sequences. Other Garmin units typically have some of these special modes as well although the key sequences may be different.

- Page - Holding the page key down while powering up the unit will cause an immediate forced cold start. On the street pilot this is the only way to observe the software revision level. On the G-III this is reported to lose the temperature compensation data.
- Mark - Holding the mark key down while powering up will totally reset the unit. You will lose all user supplied data and preferences. The machine will be set back to factory defaults. **Be careful, there is no warning message. All data will be lost!** To continue go to the section called 'after a reset'.

On some units this feature has been documented in the manual and will offer a warning message before erasing. However, some versions of software, even on these machines, do not offer the warning. Do not depend on the documentation here. Have everything of value backed up before trying this. The big plus of this feature is that this reset can fix problems with the unit that will avoid having to send it back to Garmin.

Note that setting back to factory defaults means everything. You will not only lose things that you can backup but also settings that you cannot. For example the user defined datum, the user defined grid, any preferences and datums that you have set up, any customization of any kind, and the leap second data will be gone. The Garmin waypoint will reappear if you have erased it. You will not have any of the tuning that was performed to calibrate your unit at Garmin so expect poor initial lockup performance. You will need to have a clear view of the sky and re-collect a full almanac. This takes about 15 minutes.

- Enter - Holding the enter key down while powering up the unit will cause a test mode screen to appear. This test screen is used at Garmin in final testing and calibration of the unit. Warning! Do not use this screen if your unit can get a lock onto satellites. It is possible that a real satellite may spoof the test mode into recalibrating the unit with the wrong data. No permanent damage will be done but you may experience a little longer lockup times or may even have to do a total cold start to get it running again. You may also experience continued longer lockup times for awhile while the unit re-calibrates itself under use. If your unit has a removable antenna then unplugging the antenna is a good way to ensure that no lock can be obtained.

The test mode screen can appear automatically if the unit detects a failure during power up. You can use this mode to verify certain operations of the unit. For example hitting each key will cause the corresponding key in the display to darken. Hitting the enter key twice in a row (on some units it is the page key) will cause a graphic pixel test which could highlight any bad pixels in your display. Hitting the same key again will further test the display. Hitting the key one more time will return to the main test screen. On units that use the page key to perform this test you can use the quit key to perform the graphic test backwards. The power/lamp key will show both an indication and actually light the lamp.

Other displayed entries are specific to internal tests performed in final test but the temperature (in Celsius), the internal and external battery voltages, and the GPS time can be useful. On the G-III+ and 12Map this mode will also display the version number of the software which has been removed from the start up screen. On some units status of the power on diagnostics can be viewed here.

While not obvious to an observer the test mode also starts the interface to emit PVT solutions in Garmin mode on units that support PVT output. Once started, this mode will continue even after a power off until you change it in the interface section by selecting "none".

Emap Startup modes

The Emap supports the same 3 startup modes as other Garmin handhelds but some of the keys are different.

- Holding enter while powering up enters the test mode as described above. The temperature display is only in Celsius and no external voltage is shown. You can run the visual screen diagnostics by holding down the esc key and then tap it to walk through all the screens.
- Holding down the esc key while powering up will reset the entire unit and you will lose all of your waypoints, routes, and tracklogs. There is a warning screen for this startup mode. To continue, goto the after a reset section below.
- Holding down the Find key and powering up forces the autolocate mode. It also places a couple of numbers in the upper right corner of the screen, but the interpretation of these numbers in unknown.

Etrex Startup modes

The etrex also has secret modes for test purposes. Hold the "up" and "page" keys and then press the "on/off" button to enter this screen. It behaves similar to the test mode described above and shows some information that is useful to the user. Note that is has been reported that the ROM test may show a false failure on this screen since a factory tester is not attached. You can run the key tests by

pressing each of the keys. You can use the page key to cycle through all of the display tests. The screen shows the status of the power on diagnostics plus internal battery voltage and external if present. A clock display shows seconds, the revision level of the software is shown, and a thermometer reading shows the internal temperature in degrees Celsius.

Hold the "up" and "enter" keys and then press the "on/off" button to reset the entire machine. You will receive a warning prompt. You can avoid this message using the sequence "Page" plus "Enter" plus "Up" and then press the "on/off" button. Hold the keys down for 5 seconds to perform the reset.

The newer etrex units (venture, legend, and vista) with the "Click Stick" use different keys to accomplish these tasks but the idea is the same. Holding the click stick down and then powering up will enter the test screen while holding the "page" key down and the "click stick" down and then powering up will reset the unit. You will be prompted with an "are you sure?" message. On these units holding down the "find" button in addition to the others will avoid the "are you sure" message. Hold the buttons for 5 seconds to clear the unit.

Note that the extra key sequence may be an even bigger master reset but this is not clear to me. A master reset may be needed to clear the unit if you have left the batteries out for a long period of time and the unit comes up with the wrong year. Later versions of the firmware have supposedly addressed this problem. To continue, goto the after a reset section below.

Garmin V

To totally reset this unit hold the Zoom Out and Quit buttons down and then press the power on. All user data will be lost as will the almanac. To continue, goto the after a reset section below.

76/Map 76

The reported method to completely reset one of these units is to hold the "menu" key and the "quit" key down and then press the "power on" key. Hold for a full 7 seconds and then release. As with the other units all information will be lost and you will need to reload the almanac and all your saved user data to make the unit usable.

For the 76S try this sequence to master reset it. Press and hold quit, Menu, and the rocker down key. Then press and release the power on key. Release the rocker down and wait for the welcome display. Now release the quit and menu keys.

After a reset

If you have to perform a full reset using one of the sequences shown above you will need to allow the unit to collect a full almanac before normal can be resumed. First a fix will need to be obtained which can take from 5 to 10 minutes of leaving your unit outdoors, stationary, with a good sky view. After the first lock you will need another 12.5 to 15 minutes to get the almanac reloaded. Check the date to insure that the information was reloaded properly. Time will be shown as UTC time since your local time offset was lost in the reset.

Re-establish all of your preferences such as time zone, daylight savings flag, etc. Reload all your waypoints and saved routes from your backup. You do have a backup don't you? WAAS capable receivers will have to re-establish the WAAS almanac as well.

Diagnostic Mode

This may be the most useful of the secret modes available on some of the older Garmin handhelds and, for some users, the hardest to access. Basically you start the Garmin normally by pressing the power button and while the opening screen is being displayed you must press each of the 4 arrow keys once in any order. If you accomplish this feat you will be rewarded with a quick switch to the status screen without waiting for the rest of the time-out. You should notice a -.- just under the satellite display to confirm your success. If it doesn't work then power down and try again. It takes a certain rhythm to be successful. This will not work on G-III or G-III+ units but most of the information is available on these units in normal mode. On G-II and G-II+ units this method will only work if you have the unit set in portrait mode. The etrex and emap do not support this mode.

The beauty of this mode is that **you can leave the unit in this mode while using it**. It adds a new Diagnostic menu item and more information in some of the displays.

- The -.- information on the status screen will be replaced with a HDOP setting once you have a fix. (HDOP stands for Horizontal Dilution of Precision and is a measure of the suitability of the geometry of available satellites to produce an accurate fix. Numbers below 2.0 mean that the fix is pretty good.) This is one of the factors used to calculate the estimated position error (epe) that appears at the top of the page. The EPE number is proprietary to Garmin but the HDOP number should be similar to one obtained for any unit since it only depends on the satellite geometry.

- Really old multiplex units may notice another difference on the status page. On some units the status bars are solid all of the time even when ephemeris data is not yet valid. On these units the hollow bars will appear only in Diagnostic mode. Unfortunately the new 12CX also has only solid bars and this mode doesn't fix them.
- The position page also has a new entry down near the clock. This is the current internal temperature in Fahrenheit degrees. On some units if you have set your preferences to metric this will be displayed in Celsius.
- Finally the menu page will have a new entry at the bottom. This "D" can be selected to enter the diagnostic page.

The diagnostic page content will vary depending on which unit you have but will generally contain diagnostic pass/fail messages at the top of the screen, the elapsed time for the unit, internal software information, and finally battery voltage.

- On some units the elapsed time counter can be reset by hitting the enter key providing an additional timer for the unit. On units that already have an elapsed time meter in the normal position page then the one here cannot be reset with the enter key.
- On some units hitting the page key will provide information about the status of the last several shutdowns.

Pressing quit will return you to the normal screens of the GPS but but you should realize that the GPS continues to work fine while you are looking at the diagnostic screen. If you hit goto or mark you will perform these functions.

It is worth mentioning a little about the elapsed time meter. Generally this is expected to be the time since the last factory reset. However, it will be reset by the power-on/mark key sequence described above and some software upgrades may reset it.

On some units the battery voltage includes both internal and external voltages. The internal voltage is indicated in .03 volt increments and seems to have this accuracy. The external voltage is indicated with the similar precision but does not have this accuracy. The external voltage has the following decimal setting for each whole voltage setting: .14, .29, .43, .57, .72, .86, and .00.

Other Easter eggs

On some units, mostly older multiplex units, you can redisplay the world globe that appeared on the opening screen. Change to the map page and then hit the power-off button but stop before holding it down long enough to power the unit down. The globe will appear and you can control the speed of its spinning using the arrow keys.

The III+ can display some icons that are not available on the icon menu. These are usually found in a downloaded map. For example the geographic place names are not available except when shown on the map. However you can use a trick to get these icons into your own waypoints. Find an entry using one of these waypoints on the screen and then press enter to convert it to a waypoint. You can now edit the waypoint values and name to be whatever you wish. These new icons will probably not be preserved when saved on a pc and re-entered since the pc program may not support them.

Internal Diagnostic reports

On the etrex models with the "click stick" you can reach an internal diagnostic message page which reports stack data and shutdown data. First change to the trip computer page and press the "zoom out" key, the "zoom in" key, and the "zoom out" key once again. This will turn on the diagnostic reports page.

Chapter 15

Working with the Interface

Introduction

All Garmin GPS receivers support a computer interface. This can be used to backup the waypoints and routes on a computer and to provide real time display information on a computer screen. In addition Garmin supports differential GPS input signals. The modes supported by Garmin receivers are given in the table below.

All of the Interface Modes

Table 2

Interface Mode	38-40-45XL	12-12XL-48	III-III+	Emap	Etrex	Notes
None	Y	Y	Y	Y	Y	
Garmin	Y	Y	Y	Y	Y	AKA GRMN/GRMN
None/NMEA	Y	Y	Y	Y	Y	NMEA out 0183 2.0
NMEA/NMEA	Y	Y	Y	Y	N	NMEA in/out
Garmin DGPS	Y	Y	Y	Y	Y	Has Garmin Interface control
RTCM/NMEA	Y	Y	Y	Y	Y	
RTCM/None	Y	Y	Y	Y	Y	DGPS input only
Text Out	N	N	III+, 12Map	Y	Y	1200 2400 4800 9600
RTCM/Text	N	N	III+, 12Map	Y	Y	
NMEA 0180, 0182, 0183 1.5	Y	Y	N	N	N	

All except the RTCM modes are covered in this chapter. RTCM is used by DGPS, which has its own chapter. Be sure that you have selected the correct mode and baud rate for the program or unit you are trying to interface with. This is the main problem with interface failures. For all moving map programs you will likely need NMEA mode with the baud rate set to 4800. For programs that upload and download data you should probably be in Garmin mode with the baud rate set to 9600. Some digital cameras will need text mode with a baud rate of 9600. In all cases set the data width to 8, no parity, and 1 stop bit. Make sure the interface mode is selected in the program and the correct COM port is selected.

Hardware Connection

The hardware interface for Garmin units meets the NMEA requirements and is sufficient to drive 3 NMEA loads. It is also compatible with most computer serial ports using RS232 protocols. The interface speed can be adjusted as needed by the particular interface but it usually set automatically to the appropriate setting when the interface selection is made. There is only a data in and data out line with ground. There are are no handshake lines nor should you attempt to set up a software handshake using xon/xoff as the unit does not recognize this and it may interfere with binary data uploads and downloads.

In order to use the hardware interface you will need a cable. See the accessories chapter for the available cables. In some dedicated applications you may need to wire your own or perhaps you would just prefer to do that. The Garmin cable connector shown in figure 27 will work for all of the handheld GPS units except the etrex and emap. A drawing for the etrex and emap is shown in the accessories chapter.

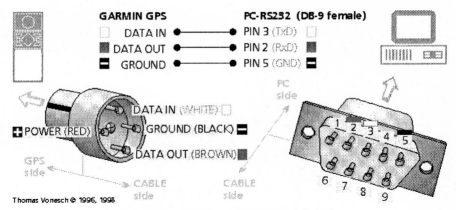

Figure 27 Cable Assembly

Garmin GPS receivers may be used to interface with other NMEA devices such as autopilots, fishfinders, or even another GPS receivers. They can also listen to Differential Beacon Receivers that can send data using the RTCM SC-104 standard.

Some of the latest computers no longer include a serial port but only a USB port. Garmin receivers are known to work with Serial to USB adapters and serial ports attached via the pcmcia (pc card) adapter.

NMEA

The National Marine Electronics Association has developed a specification that defines the interface between various pieces of marine electronic equipment. The standard permits marine electronics to send information to computers and to other marine equipment. GPS receiver communication is defined within this specification. Most computer programs that provide real time position information understand and expect data to be in NMEA format. This data includes the complete PVT (position, velocity, time) solution computed by the GPS receiver. The idea of NMEA is to send a line of data called a sentence that is totally self contained and independent from other sentences. There are standard sentences for each device category and there is also the ability to define proprietary sentences for use by the individual company. All of the standard sentences have a two letter prefix that defines the device that uses that sentence type. For GPS receivers the prefix is GP. This is followed by a three letter sequence that defines the sentence contents. In addition NMEA permits hardware manufactures to define their own proprietary sentences for whatever purpose they see fit. All proprietary sentences begin with the letter P and are followed with a letter that identifies the manufacturer controlling that sentence. For Garmin this would be a G.

Each sentence begins with a '$' and ends with a carriage return/line feed sequence. The data is contained within this single line with data items separated by a comma. The data itself is just ascii text and may extend over multiple sentences in certain specialized instances but is normally fully contained in one variable length sentence. An example sentence might look like:

```
$GPGGA,123519,4807.038,N,01131.000,E,1,08,0.9,545.4,M,46
.9,M,,*42
```

```
With an interpretation of:
```

```
      GGA - Global Positioning System Fix Data
      123519       Fix taken at 12:35:19 UTC
      4807.038,N   Latitude 48 deg 07.038' N
      01131.000,E  Longitude 11 deg 31.000' E
      1            Fix quality: 0 = invalid
                                1 = GPS fix
                                2 = DGPS fix
      08           Number of satellites being tracked
      0.9          Horizontal dilution of position
      545.4,M      Altitude, Meters, above mean sea level
      46.9,M       Height of geoid (MSL) above WGS 84
                        ellipsoid
      (empty field) time in seconds since last DGPS update
      (empty field) DGPS station ID number
      *42          the checksum data, always begins with *
```

Each Data type would have its own interpretation, which is defined in the NMEA standard. This particular sentence provides essential fix data. Other sentences may repeat some of the same information but will also supply new data. Whatever is reading the data can watch for the data sentence that it is interested in and simply ignore whatever sentences that is doesn't care about. In the NMEA standard there are no commands to indicate that the GPS should do something different. Instead each receiver just sends all of the data and expects much of it to be ignored. On NMEA input the receiver stores information based on interpreting the sentence itself. While some Garmin receivers accept NMEA input this can only be used to update a waypoint or similar task and not to send a command to the unit. There is no way to indicate whether the sentence is being read correctly or to re-send some data you didn't get. Instead the receiving unit just checks the checksum and ignores the data if the checksum isn't correct figuring it will be sent again sometime later. No error can be generated to the remote system.

The NMEA standard has been around for many years and has undergone several revisions. The protocol has changed and the number and types of sentences may be different depending on the revision. All Garmin receivers understand the latest standard which is called: 0183 version 2.0. This standard dictates a transfer rate of 4800 baud. Some Garmin receivers also understand an earlier version of 0183 called version 1.5. Some Garmin receivers even understand older standards. The oldest standard was 0180 followed by 0182, which transferred data at 1200 baud and had very few sentences defined. Some Garmin units can be set to 9600 for NMEA output but this is only recommended if you have determined that 4800 works ok and then you can try to set it faster.

If you are interfacing a Garmin unit to another device, including a computer program, you need to insure that the receiving unit is given all of the sentences that it needs. If it needs a sentence that Garmin does not send then the interface to that unit is likely to fail. The sentences sent by Garmin receivers include:

NMEA 2.0
GPBOD bearing, origin to destination - earlier G-12's do not transmit this
GPGGA fix data
GPGLL Lat/Lon data - earlier G-12's do not transmit this
GPGSA overall satellite reception data
GPGSV detailed satellite data
GPRMB minimum recommended data when following a route
GPRMC minimum recommended data
GPRTE route data
GPWPL waypoint data (this is bidirectional)

NMEA 1.5 - some units do not support version 1.5

GPBOD bearing origin to destination - earlier G-12's do not send this

GPBWC bearing to waypoint using great circle route.

GPGLL lat/lon - earlier G-12's do not send this

GPRMC minimum recommended data

GPRMB minimum recommended data when following a route

GPVTG vector track and speed over ground

GPWPL waypoint data (only when active goto)

GPXTE cross track error

In addition Garmin receivers send the following Proprietary Sentences:

PGRME (estimated error) - not sent if set to 0183 1.5

PGRMM (map datum)

PGRMZ (altitude)

PSLIB (beacon receiver control)

The etrex summit, Vista, and 76S send a $HCHDG sentence for their compass output.

This list is specific to the handheld units. Other Garmin units may send other sentences and some use proprietary sentences to send control commands to the units themselves. Note that Garmin converts lat/lon coordinates to the datum chosen by the user when sending this data. This is indicated in the proprietary sentence PGRMM. This can help programs that use maps with other datums but is not a NMEA standard. Be sure and set your datum to WGS84 when communicating to other NMEA devices.

It is possible to just log view the information presented on the NMEA interface using a simple terminal program. If the terminal program can log the session then you can build a history of the entire session into a file. More sophisticated logging programs can filter the messages to only certain sentences or only collect sentences at prescribed intervals. Some computer programs that provide real time display and logging actually save the log in an ASCII format that can be viewed with a text editor or used independently from the program that generated it.

Some handhelds such as the Map76S can change the sentences for NMEA mode. A Setup NMEA output item is on the Interface setup page local menu when the Serial Data Format is set to NMEA. This allows setting the following for the NMEA output: the precision of the minutes field for lat/lon, the waypoint identifier format, and which groups of sentences to transmit. These settings allow compatibility with a wider range of NMEA-driven products including some older marine autopilots.

NMEA has its own version of essential GPS pvt (position, velocity, time) data. It is called RMC, The Recommended Minimum, which might look like:

```
$GPRMC,123519,A,4807.038,N,01131.000,E,022.4,084.4,23039
4,003.1,W*43
```

```
With an interpretation of:
```

RMC	Recommended Minimum sentence C
123519	Fix taken at 12:35:19 UTC
A	Status A=active V=Void
4807.038,N	Latitude 48 deg 07.038' N
01131.000,E	Longitude 11 deg 31.000' E
022.4	Speed over the ground in knots
084.4	Track angle in degrees True
230394	Date - 23rd of March 1994
003.1,W	Magnetic Variation
*43	The checksum data, always begins with *

NMEA input

Many of the Garmin units also support an NMEA input mode. While not too many programs support this mode it does provide a standardized way to update or add waypoint and route data. In addition the 76 family uses this mode to input depth data from an NMEA compatible depth finder. Note that there is no handshaking or commands in NMEA mode so you just send the data in the correct sentence and the unit will accept the data and add or overwrite the information in memory. If the waypoint name is the same you will overwrite existing data but no warning will be issued. The sentence construction is identical to what the unit downloads so you can, for example, capture a WPL sentence from one unit and then send that same sentence to another unit. But, be careful if the two units support waypoint names of different lengths since the receiving unit might truncate the name and overwrite a waypoint accidentally. If you create a sentence from scratch you should create a correct checksum. Remember that NMEA requires both a carriage return and a line feed at the end of the line. Be sure you know and have set your unit to the correct datum. A WPL sentence looks like:

```
$GPWPL,4807.038,N,01131.000,E,WPTNME*31
```

```
With an interpretation of:
```

```
WPL            Waypoint Location
4807.038,N     Latitude
01131.000,E Longitude
WPTNME         Waypoint Name
*31            The checksum data, always begins with *
```

Garmin mode

Garmin mode is a bi-directional binary proprietary interface protocol that is used by Garmin and many third party vendors do communicate directly with a Garmin receiver. All of the handheld units understand Garmin protocol but may not understand or respond to a specific command in that protocol. For example, if you were to try and store altitude information in a waypoint on a unit that cannot store altitude information then this command would fail. Garmin mode includes a set of published API (Application Programming Interface) specifications and other commands that are not published or made public in any fashion except as used by a Garmin product. There are other commands that are not even used by a Garmin product and are probably used by internal test groups and custom test equipment at the Garmin factory. It is beyond the scope of this manual to describe the detailed interface specification. A manual describing the published interface specifications is available for downloading at the Garmin site. In addition some of the undocumented commands and features are available from web sources.

Some of the things that you might be able to do using Garmin protocol include:

1. getting the version number of the software.
2. Finding out the capabilities of the unit.

3. Uploading or downloading waypoint data.
4. Uploading or downloading Track data.
5. Uploading or downloading Route data.
6. Uploading or downloading Almanac data.
7. Downloading the current GPS time.
8. Downloading the current GPS position.
9. Uploading a new release of the firmware.
10. Uploading a set of maps.
11. Downloading a screen snapshot.
12. Receive a complete PVT solution in real time.

Older multiplexing units cannot do many of the items in the list. Specifically they cannot do item 2, or any item beyond 8 in the above list. The G-12 family cannot do items above 9 but this could change with new firmware releases.

User data backup

The most often used capabilities include the backup of critical user data such as waypoints and routes. To do this you would need to secure a Garmin capable program. These are available from Garmin or several third party sources. There are programs for pc's running dos, or windows, for Macs, for Unix, for palms, and pocket pc's. Once you have the correct program you can place your unit in Garmin mode and set the baud rate to 9600. Generally all programs accept this baud rate but some may support other rates and even change the rate. Cable the unit to the computer and make sure the computer program is set to the correct serial port and the baud rate to 9600. The standard serial port settings are 8N1, 8 bit data, no parity, 1 stop bit. Do not use xon/xoff since this may interfere with proper transfer of binary data. If the program cannot access the unit then check to ensure some other program is not using the port (the palm sync program is notorious for this) and that the port is configured correctly. This protocol is binary and requires handshaking so all three wires need to be hooked up correctly. Perhaps a null model adapter may be required to get the receive and transmit signals hooked up properly.

You cannot modify individual waypoints or routes using this interface. Instead you load the full set of routes or the full set of waypoints. If you wish to revise certain data you should download the full set and then revise the data you wish on your computer, clear all the waypoint data, and then reload the full set back. Otherwise you may get unpredictable results on some units. Most units will simply overwrite waypoints with the same name but the emap will create a new waypoint if the location is different. This can make updating a waypoint a bit frustrating. Similarly you load the full tracklog. On units with multiple tracklogs you may find them all concatenated together on download. Some programs may be able to upload saved logs directly but some may not. On units that support the uploading of maps the rule is similar. You must assemble all of the maps you wish to upload and then send them all at once replacing the previous upload.

Of course, the data in the computer program need not originate in your GPS. It is quite possible, in some programs, to import external data to the program for later uploading or to edit the data files directly to provide this new information. It is also useful to modify and add comment data to waypoints using a computer keyboard instead of toggling it in with the unit keypad. Be aware that waypoints are always interpreted as using the WGS-84 datum for this interface.

Unit to Unit transfer

One interesting use for this protocol is to transfer information between units. To do this you need a Garmin to Garmin cable (available from Garmin and other sources) to hook the two units together. One of the units is placed in Host mode and the other unit sends commands to upload and download data. The commands are shown on your units menu. They may include:

1. RQST/SEND ALL USR
2. RQST/SEND CFG - configuration
3. RQST/SEND PRX - Proximity alarm data
4. RQST/SEND RTE - Routes
5. RQST/SEND TRK - Track Log
6. RQST/SEND WPT - Waypoints
7. RQST/SEND ALM - Almanac

The emap and etrex do not support this mode. Other units may not work correctly in this mode or may not support some commands. For example a G-III does not have proximity alarms. The G-38 and the G-12 can talk to each other but the G-38 will not be able to support waypoint icons and you can easily overflow the track log on a G-38 with the tracklog on a G-12. It is also possible to use a computer as an intermediary for this transfer by moving the data from the unit to the computer and then uploading it to the target platform. Some Palm pilot programs even support the host mode so that they can serve as a temporary storage point for GPS user data.

Firmware Upgrade

Garmin releases firmware upgrades for all of their twelve channel units to fix bugs and to add functionality. These upgrades are available from the Garmin web site and are free so long as you agree with the terms and conditions. They come with the appropriate program for pc platforms and are only supported by Garmin. There is no third party source and users on Macintosh units will have to find a friend to do the upgrade or use a pc emulator. Be careful that you only try to use the firmware for your specific unit, or you could break your unit completely and have to send it back to Garmin for repair. The older multiplex units cannot be upgraded in this fashion and if needed they must be returned to Garmin for any upgrades. Be sure to read the instructions that accompany the upgrade at the Garmin site.

To ensure success, make sure you download the upgrade using a binary mode. It comes as a zip file so if it unzips correctly you can be sure you downloaded it correctly. Make sure you have a good connection to the GPS. Try one of the Garmin interface programs to backup your data. Generally an upgrade does not lose user data but this is not always the case so it is a good idea to back it up. Leave the unit in Garmin mode for the upgrade. Your pc baud rate should be set to "maximum rate" possible so that the program can increase the baud rate to minimize the download time. Expect to lose any customization that you may have performed on your unit. Be sure you have fresh batteries in the unit. Writing the firmware into the prom can use significant battery power and if the batteries are weak you may not get a good load or you may start out with a seemingly good load that will fail later. Do not abort the process once it has begun. It can take several minutes to do the upgrade so be patient. If the upgrade fails, try it again. You must get an good upgrade before your unit will be operational again.

If you lose power or connection during the upgrade you may have a unit without any code at all. If you were to attempt to power up the unit it would tell you the firmware is missing. Some users have reported that this happened some time later. This is usually caused by weak batteries that were not able to burn the new code in the machine. To recover perform these steps:

1. Connect your cable to the computer.
2. Get the computer ready to load the firmware but don't press ENTER.
3. Turn on the GPS and then press the ENTER key on your computer.
4. Watch - The program should begin to load the new firmware

If you are unsuccessful then call Garmin and arrange to return your unit for them to upgrade. If you feel that the upgrade has a bug in it and the older release is better you can generally use an earlier upgrade to downgrade your unit. Garmin generally does not keep older versions available but they are often available on the net from other users.

PVT data

Some of the Garmin receivers support a PVT mode as part of the Garmin mode. If you are using a computer program that supports this then you can remain in Garmin mode even while running your real time mapping application. Set your unit to Garmin mode and then select this solution from the menus in the application. Delorme mapping products support this mode. This is an advantage in that you don't need to switch modes and you can leave your interface at 9600 baud, which makes the real time response a bit faster. The update interval is 1 second and this mode does not require handshaking nor does it support retransmission of data. The following data is typically included as part of the pvt structure in the D800 message:

alt - Altitude above WGS-84 ellipsoid
epe - total predicted error (2 sigma meters)
eph - horizontal position error
epv - vertical position error
fix - type of position fix
tow - time of week (seconds)
posn - lat/lon (radians)
east - velocity east (meters/sec)
north - velocity north (meters/sec)
up - velocity up (meters/sec)
msl_height - height of WGS-84 ellipsoid above MSL (meters)
leap_seconds - difference between GPS time and UTC (seconds)
wn_days - week number days

Undocumented modes

The Garmin Interface specification defines much of the exact interface requirements for Garmin mode. However, there are many things that are in the interface that are not described in this manual. Garmin has indicated that these are for test purposes and are not to be used by customers. They may also be changed from release to release and may only work with a particular test setup. However, many of these modes have been discovered and decoded by third party programmers. Such additional features as screen captures fall into this category. One vendor has actually managed to get the pseudo range data out of the Garmin 12 family and provides a post processing capability with these units by collecting data on a pc in real time for later processing thereby opening the possibility of using this unit for surveying applications.

Raw Data

While not documented, the G-12 family, newer etrex models, and some other Garmin models can be coerced into sending raw pseudo range data. This is the data obtained from the satellites prior to using it for a position solution and it can be used by post processing programs on a computer to correct for certain kinds of errors and permit better information about your position. This is the technique used by many Surveyors to obtain more accurate data. The idea is to hook the G-12 family unit to a computer and collect the raw data on the computer for later analysis. The G-12 cannot collect this data on its own. One source of a program is Async Logger, which can collect the data and convert it to survey industry standard RINEX2 (Receiver independent exchange) format.

The procedure to obtain this data is:

1. Power on your GPS12 with a good view of the sky.
2. Wait until a 3D fix is obtained.
3. Run the async logger for a while (let's say 5 minutes): async -p your_port -a 0xffff -t 300 -o data.g12
4. Run the parser with the option -rinex and redirect the output to a file: parser data.g12 -rinex data.RNX
5. Postprocess the RINEX file with postprocessing software.

A similar procedure will be used for the other tools. You will need to find your own postprocessing software and a source of correction data. Pointers for this data can be found at the above site. Another source for this kind of processing is Gringo, and a product called Rhino from US Positioning. Rhino is able to extract pseudo range data and take advantage of the carrier data for even better accuracy. These tools provide postprocessing software as part of the package.

Text Mode

Text mode is a simple output mode that supplies velocity and position information in realtime. Currently this is primarily used by certain digital cameras to include this data on the picture. In the future many other uses will be found for this mode which requires very little processing on the part of the device receiving the data. An example is shown below.

```
@000607204655N6012249E01107556S015+00130E0021N0018U0000
@yymmddhhmmss Latitude Longitude error Altitude EWSpd NSSpd VSpd
```

Each item is of fixed length making parsing by just counting the number of characters an easy task. It is grouped by use permitting a digital camera, for example, to just read the first 30 characters and report the time and position. Some of the data will require some programming to make it meaningful for most users, such as the speed which is divided into the X, Y, and Z components. This is the only format that provides vertical speed, which should be a great boon for balloonists.

A more formal description of the fields are shown in Table 3:

Table 3 Text Mode

FIELD	DESCRIPTION:	WIDTH	NOTES:
	Sentence start	1	Always '@'
T	Year	2	Last two digits of UTC year
I	Month	2	UTC month, "01".."12"
M	Day	2	UTC day of month, "01".."31"
E	Hour	2	UTC hour, "00".."23"
	Minute	2	UTC minute, "00".."59"
	Second	2	UTC second, "00".."59"
P	Latitude hemisphere	1	'N' or 'S'
O	Latitude position	7	WGS84 ddmmmmm, with an implied decimal after the 4th digit
S	Longitude hemisphere	1	'E' or 'W'
I	Longitude position	8	WGS84 dddmmmmm with an implied decimal after the 5th digit
T	Position status	1	'd' if current 2D differential GPS position 'D' if current 3D differential GPS position 'g' if current 2D GPS position 'G' if current 3D GPS position 'S' if simulated position '_' if invalid position
I			
O			
N	Horizontal posn error	3	EPH in meters
	Altitude sign	1	'+' or '-'
	Altitude	5	Height above or below mean sea level in meters
V	East/West velocity – direction	1	'E' or 'W'
E	East/West velocity – magnitude	4	Meters per second in tenths, ("1234" = 123.4 m/s)
L	North/South velocity – direction	1	'N' or 'S'
O	North/South velocity – magnitude	4	Meters per second in tenths, ("1234" = 123.4 m/s)
C	Vertical velocity – direction	1	'U' (up) or 'D' (down)
I	Vertical velocity – magnitude	4	Meters per second in hundredths, ("1234" = 12.34 m/s)
T			
Y	Sentence end	2	Carriage return, '0x0D', and line feed, '0x0A'

Notes on the table:

- If a numeric value does not fill its entire field width, the field is padded with leading '0's (eg. an altitude of 50 meters above MSL will be output as "+00050").
- Any or all of the data in the text sentence (except for the sentence start and sentence end fields) may be replaced with underscores to indicate invalid data.

Chapter 16
Differential GPS

This chapter covers the use of differential corrections on Garmin GPS receivers. DGPS is a method of improving the accuracy of your receiver by adding a reference station to augment the information available from the satellites. For traditional dgps this station transmits correction data in real time that is received by a separate box, called a beacon receiver, to send correction information to the GPS receiver. WAAS is another differential correction method that gets its corrections from a satellite. This data is processed internally by a WAAS capable GPS receiver.

Frequently Asked Questions

1   Q: What is DGPS?
   A: DGPS is a method of improving the accuracy of your receiver by adding a local reference station to augment the information available from the satellites. It also improves the integrity of the whole GPS system by identifying certain errors.

2   Q: Why would I need one?
   A: You might not, but if you want to improve accuracy beyond what is available for a consumer grade GPS receiver or want to improve integrity by knowing when to believe the receiver, one of these solutions may be right for you.

3.   Q: Just how accurate is my GPS receiver any way?
   A: Most GPS manufacturers quote 17 meters (49 feet) as the accuracy of horizontal positions anywhere on earth. However, independent testing has shown that modern receivers can achieve 10 meters fairly reliably with a clear sky view.

4. Q: But *my* receiver reports errors much less than that. Do I have an exceptional unit?

A: A GPS can only estimate the accuracy and many manufacturers are a bit optimistic in the numbers they quote. To be more precise the Accuracy number presented is often based on a 50% to 60% probability rather than the 95% probability that is usually considered in a scientific evaluation. Note that no receiver can guarantee a particular level of accuracy without stating a probability and one of the features of some of the systems described below is to help identify when the data in the GPS might have a higher than average amount of error.

5. Q: What are the sources of this error?

A: They are well understood and the dominant contributors are listed in this table along with the likely amount that they contribute. (This assumes a good sky view and reasonable satellite geometry.) Note that in real life the errors may be higher or lower than those listed in the table.

Table 4

Error	Value
Ionosphere	4 meters
Clock	2.1 meters
Ephemeris	2.1 meters
Troposphere	0.7 meters
Receiver	0.5 meters
Multipath	1 meter
Total	**10.4**

6. Q: So what improvement can I expect?

A: A beacon based DGPS system is somewhat dependent on how close you are to the beacon, but you can achieve 1 to 5 meters overall accuracy. A WAAS system can achieve an accuracy of less than 3 meters if you are located in the area where ionospheric correction data is available. These are still not as accurate as survey grade receivers that receive dual frequencies, have a DGPS reference station located very close to the survey site, and use post processing techniques to reduce errors even further.

DGPS mode using a beacon receiver

All Garmin receivers support DGPS. This station transmits correction data in real time that is received by a separate box, called a beacon receiver, which sends the correction information to the GPS receiver. In principle this is quite simple. A GPS receiver normally calculates its position by measuring the time it takes for a signal from a satellite to reach its position. By knowing where the satellite is, how long it takes to send the signal, and knowing the speed of the signal a GPS can compute what is called a pseudo range (distance) to the satellite. This range must be corrected before it is used to compute the final position. Compensation for ionospheric errors due to the fact that the ionosphere slows down the speed of travel of the radio wave is one form of correction that can be applied. A DGPS beacon transmitter site has already calculated all of the pseudo range correction data based on the fact that it already knows exactly where it is and can compute the errors in the satellite computed position from its known location. Once the pseudo range correction data is computed it is sent to the GPS and used to compute a more accurate fix. The data is sent at either 100 baud or 200 baud depending on the station and this can result in a typical delay of 2 to 5 seconds between the computation of the correction and the application of the correction. However, since most errors are slow moving this time delay is not usually a problem.

DGPS mode

Each beacon transmitter is autonomous and computes its own corrections based on its reception of GPS signals. It then packages the correction data in groups of 3 satellites and sends the data to the GPS receiver. Note that the design of a beacon DGPS transmitter will send corrections for up to 9 satellites and these are only those at least 5 degrees above the horizon. The assumption is that the GPS receiver will be close enough to have the same sort of errors that the beacon station saw and they can be applied without modification to any SV's that they share a view of. This works well in practice since most of the error sources shown above in the FAQ's would be common between the two locations. Beacon sites have some ability to improve the system integrity as well, however there is no standard that is defined as to exactly what they can identify. They can easily identify a satellite where the required corrections exceed a prescribed value and should not be used.

It seems that Garmin will favor differentially corrected satellites when at least four exist to the exclusion of regular satellites. If the four are in a poor geometric relationship the epe number, and possibly the accuracy of the solution, can be worse than it was with a regular GPS solution.

While the major source of DGPS corrections are done via beacon transmitters operating in the 300KHz band this is not the only source of correction data. It is possible to get data from any source that can be received at your location. Some sources include FM radios using the subcarrier capability of these transmitters, the Internet, and even satellites. In all cases a custom receiver (or software) is used to assemble the data in a form that is acceptable to the GPS receiver, which by standard is RTCM-104. Data conforming to this standard is then sent via the serial port to the GPS receiver on a cable.

Cable for DGPS

Most folks fabricate a custom cable to work with the beacon receiver. Here is a diagram for a fairly complicated version, but you may not need a setup that is this complicated depending on what else you may be doing.

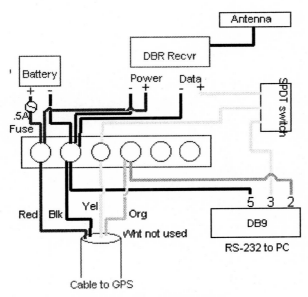

Figure 28 DGPS Cable

For simple DGPS connections you can just wire a beacon receiver output signal along with its ground to the data input terminals of the GPS. If you need to be able to control the beacon receiver from the GPS receiver then you will also need to send the output for the GPS receiver to the beacon receiver. A standard computer interface cable can usually be used for this connection. If you also need to talk to a pc at the same time things start to get a little more complicated. To talk to a pc in NMEA mode you can simply send the output of the GPS to both units. Wire the output signal to the input on the computer and the input on the beacon receiver. There is sufficient power in the signal from the Garmin to drive both units and even a third item. Note that if the beacon receiver doesn't need to receive commands from the GPS then there is no reason to send the signal both places but the ground

243

wire is still needed. Finally if the GPS needs to talk to the pc in Garmin mode and also to the beacon receiver you will need a switch to permit the beacon receiver to transmit difference signals or the pc to interact with the GPS. You won't be able to do both at the same time. This should not present any real problems since the bi-directional Garmin mode is used to upload and download waypoint, route, and track data which does not need the beacon receiver to be operational.

WAAS

WAAS, Wide Area Augmentation System, is the latest method of providing better accuracy from the GPS constellation. It is similar in principle to the DGPS capability that is built into all Garmin units except that a second receiver is not required. Instead of a beacon receiver the correction data is sent via a geo-stationary satellite and is decoded by one of the regular channels already present in the GPS receiver. Thus one of the 12 channels can be designated to decode regular GPS signals or can be used to decode the WAAS data. Actually, as currently implemented, when WAAS is enabled two channels will be dedicated to WAAS. While WAAS is the name of the implementation of this technology in the US the system is intended for worldwide use. All of the newer handheld units are capable of receiving WAAS signals including the etrex Venture, Legend, and Vista models, the 76 family, and the GPS V.

The way this works is that a set of ground stations all over the US (as shown below) collect correction data relative to the area of the country where are they are located. The entire data is then packaged together, analyzed, converted to a set of correction data by a master station and then uploaded to the geo-stationary satellite, which in turn transmits the data down to the local GPS receiver. The GPS receiver then figures out which data is applicable to its current location and then applies the appropriate corrections to the receiver. Similar systems are being set up in other areas of the world but they are not yet operational.

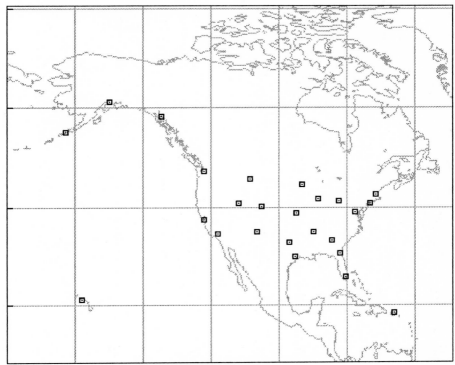

Figure 29 WAAS map

In addition to correction information the ground stations can also identify a satellite that is not working within specification thereby improving the integrity of the system for aviation use.

As stated, the intent is to have worldwide coverage of WAAS corrections, however the name of the correction system varies. In Europe it is called EGNOS while in Asia the Japanese system is called MTSAT, but whatever it is called the system is designed to be compatible worldwide through a cooperative effort in member countries. The European ground station network is shown below. Europe is currently in test mode so the Garmin unit will not use the corrections.

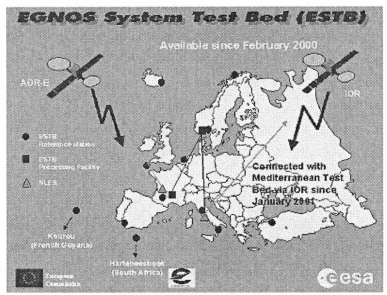

Figure 30 EGNOS map

The Garmin unit identifies these geo-stationary satellites on the satellite status screen by using numbers greater than 32. This system is just being set up now and will be improved with more satellites in the future (possibly 19 of them world wide), however since they are all geo-stationary you will need a clear view of the southern sky to use them from the northern hemisphere. This means they are very useful for an airplane or perhaps a boat, but less useful to someone on the ground particularly in areas of tree cover or high northern latitudes. While a GPS receiver can possibly receive satellite data from outside the ground coverage area there will be very little correction capability without the correct ground data.

Loading the Almanac

The latest released products from Garmin include WAAS capability (called EGNOS in Europe and MSAS in Japan). For Garmin this includes the etrex: Venture, Legend, and Vista models as well as the GPSMap76, GPS76, and Garmin V. They have also updated some other units with this capability such as the aviation units like the GPS 295. However, unlike the standard GPS almanac that is preloaded into each Garmin receiver, the WAAS almanac is not loaded into the receiver when you get it. Each person is required to get the almanac for himself or herself and this step is required before the GPS can be used in WAAS mode. Some folks are able to get an almanac fairly quickly while others struggle for days and are still not able to get a successful load. Here is the technique that will result in obtaining the almanac in the minimum amount of time.

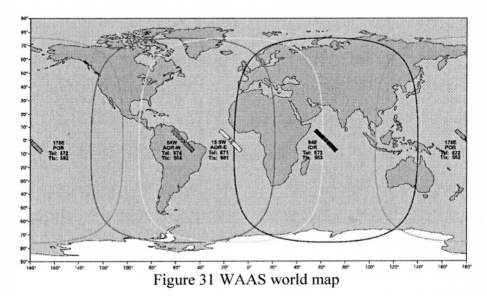

Figure 31 WAAS world map

1. Study the map of WAAS GEO satellites and determine which ones you are going to be able to see from your location. From the west coast of the US you can access number 47 and number 35 while from the east coast you can only see number 35. (Also the EGNOS 33 of course.)
2. Find a place with a clear sky view in the direction of the WAAS GEO satellites you are interested in. If necessary set a waypoint to their approximate location so that you can use the GPS itself to provide a bearing to the satellite. GEOS 35 is at Lat 0 and Lon 54 West while 47 is at Lat 0 and Lon 179 West. (The Europe EGNOS system satellite is at Lat 0 Lon 15.5 West.)
3. Be sure that your GPS is set to normal mode (not battery save) for this procedure. The nature of WAAS corrections precludes its use in Garmin battery save mode.

4. Enable WAAS on your GPS and notice that it takes the last two locations on the satellite page to display WAAS Geo activity. It will cycle through all of the 19 possible satellite locations 2 at a time and then repeats until two candidate satellites are found. It stays at one setting for about 45 to 50 seconds and then selects a new pair of SV's to look for. When it doesn't have a loaded almanac it will show the SV's on top of the N indicator on the satellite page.

5. When it reaches the ones you know it should see you need to make sure that the SV shows some receive strength. It it doesn't then reposition the unit in an attempt to find either a location or a direction that will cause the signal bar to appear. You only have 45 seconds so you need to try several positions to get that signal bar. Point directly at the SV, change the angle slightly (up to 90 degrees) and tilt the antenna in the direction of the SV, leave it in each trial position for a few seconds and if not successful then try another setting. You may need to move a little. The tree you thought was out of the way may be blocking the signal.

6. Once the satellite strength bar appears you are home free. It will not move off that SV any longer although the second position will continue to hunt for another SV unless it too gets a signal bar. You can reposition slightly to help the second one lock on as well but be careful that you don't lose the first one.

7. Hold the unit until the full Almanac gets loaded. This will be indicated by the satellite(s) assuming the correct position on the page and will usually turn solid (but this is not a requirement). The figure below shows two GEO SV's in position from a location in California. When the satellite is in position and collecting data it will also begin working by starting to display small D's on the other satellite bars to indicate that they are in differential mode. It can take up to 5 minutes to load the almanac data and if it misses any of the data you could wait another 5 minutes for it to come around again. If it grabs a SV that does not have land data for your area it could take quite a while for

it to figure this out, but it will eventually supply some amount of differential correction.

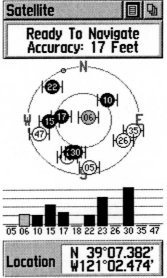

Figure 32 etrex with WAAS SV's

8. If the unit seems to ignore a satellite you know it should see and then switches and moves on to other satellites without generating a signal bar then you need to re-evaluate your location and find a better spot with fewer in the direction of the satellite. You will need to wait until it gets back to your satellite location before trying again. If you are searching for a low number you may speed up the search by resetting the search to begin at the beginning by turning off the unit or disabling and re-enabling WAAS mode. Note that during this early phase of WAAS/EGNOS/MSAS it is likely that the system may be in test mode and not be supplying consistent data. Therefore you may see delays in lock and outages like the signal suddenly being reset, however in the US these outages are rapidly becoming a thing of the past since the system has settled down and is considered to be nearly operational.

9. Once you have one or two SV's locked you are ready to use WAAS mode. Congratulations. The unit knows

whether it is in a location where there are one or two satellites available so if you collect the almanac and the unit determines only one satellite is available it will give the second channel back to the GPS for normal GPS satellite use.

Well, now that you have the almanac data, you might be interested in what you just loaded. For a GPS the almanac data includes a coarse position data for all possible GPS SV's. Similarly the GEO almanac includes coarse position data for all possible 19 GEO satellites including the fact of their existence. The almanac will be updated as new satellites are launched but will otherwise remain static since these birds don't move much. The UTC time will also be loaded showing when the data was collected, and an ionospheric grid mask will be loaded (described under how WAAS works). While locked the unit will also collect ionospheric correction data, ephemeris correction data, clock correction data, integrity data and everything else since everything repeats at least once in 5 minutes.

Technical details on how WAAS works

Similar to a beacon transmitter the WAAS system collects data from strategically placed and well characterized ground locations. However, unlike the standard dGPS a WAAS system cannot directly send corrections to the pseudorange data. This is because the unit has no idea where you are. Instead it attacks the problem by addressing the individual sources of error and sending corrections for each one of them. If you remember from the FAQ on error sources the biggest component of error is due to ionospheric delays followed by clock errors and ephemeris errors. In addition another significant error source was troposphere errors. To attack these error sources the WAAS system sends clock corrections, ephemeris corrections, and ionospheric corrections. It cannot compute tropospheric corrections due to the localized nature of this error, but it does remove the tropospheric error component from the data it computes so that the local receiver can apply its own corrections. This would be based on an atmospheric model that is calculated with reference to the current sky location of the SV. The ground stations do not send the data directly to a WAAS satellite for re-transmission but rather to a master ground station that analyzes the input and computes the full detailed error information. This fully correct data is then sent to the GEO satellite to be sent to the GPS receiver. Note that there is no correction done by the GEO satellite; the master ground station will even correct for errors produced by the GEO satellite itself. There are redundant master stations as well for system wide integrity.

Clock errors can change rapidly so this data is update every minute if required. Ephemeris errors and ionosphere errors don't change nearly so fast so they are only updated every 2 minutes and can be generally be considered valid for up to 3 times that period of time. Even this time is very conservative in practice. Clock and ephemeris data is specific to a satellite but ionospheric errors are specific to your location therefore they must be sent separately. After receiving raw data from each of the ground station the master station divides the country into a grid and then builds ionospheric correction information on a per grid location basis from the data received from each reporting station. It is this grid location that is used by the GPS receiver to determine the applicable ionospheric corrections. In addition the master station determines the validity of the data it receives and can indicate invalid data within 6 seconds to the GPS receiver.

Of the forms of error correction supported by WAAS only the ionospheric data requires knowledge of the receiver position. Clock and ephemeris data is available for any receiver even if it is currently located outside the area covered by ionospheric correction data. In addition system integrity data can be used outside this area. The designers of WAAS developed a grid system of correction data that permits a receiver to use the data it needs for this correction. Here is what the grid system looks like.

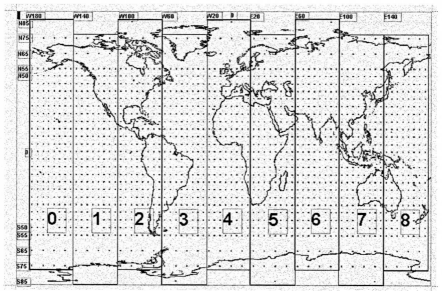

Figure 33 WAAS Grid

The master ground station computes a correction for ionospheric data for each of the points on the grid based on field data it gets from the other ground units. Of course it is very unlikely that there will be data for all of the points on the grid over the entire earth so the GPS receiver downloads a grid mask that is part of the almanac data that tells it where to expect corrections. The mask is divided into bands as shown on the drawing and each band contains 201 grid points (except the last one that has 200). A single bit is used to represent the availability of data on each grid point so the entire band can be contained in a single packet of data. Each geo satellite can have data for up to a maximum of three or four bands but may have less. The master station will generate correction data or each of the grid points tagged in the mask. A receiver will locate its position relative to 4 grid points and interpolate the data from those 4 points based on their relative distance. If only 3 points have data then the receiver will compute a triangle of the 3 points and if it is inside the triangle it will use the 3 points to interpolate its correction. Otherwise, the use of any of the grid points for correction is undefined. It seems however that most GPS implementations will use the data from even a single point if it is "close" to the current location. As can be seen from the map of the USA, phase one implementation, this is required to provide

252

WAAS for areas such as Alaska and Hawaii. It may be that multiple grid locations are generated from the single data input.

In addition to correction data the WAAS system places a high degree of importance to system integrity. Each ground station and the master station has independent sources of critical data and can determine if an SV is out of calibration. Bad data can be identified and relayed to the receiver within about 6 seconds. The geosynchronous satellites that retransmit the data can even be used as regular GPS satellites as part of the regular GPS solution since they also relay regular satellite ephemeris data for themselves. This data, like all of the other data, is generated at the master station and can be turned off independent from the regular WAAS corrections if the satellite drifts too far.

All of the data for a region is loaded into the WAAS SV so only one is required to receive everything. A second provides redundancy.

How WAAS gets used by the Garmin receiver

The information in this section is strictly conjecture based on study and observation. Garmin considers the inner-workings of their GPS receivers to be trade secrets so this information is not published. Note that WAAS use requires the GPS be in normal mode. Battery-save mode does not lend itself to the intermittent processing capability of battery-save mode. Generally the SV's for WAAS won't even show up in battery-save mode but it is possible to get the unit into battery-save and have it still attempt WAAS processing but the results are unpredictable and generally not usable.

It is possible to leave the WAAS enabled all the time. The only negative is that the unit will expend some energy attempting to lock onto the WAAS satellites and the two channels will not be available for standard GPS devices. Even a momentary break in terrain will cause the GPS to lock onto a GEO SV if it can see one. WAAS data is sent in packets that are one second long (250 bits) and a lock can occur at any one second interval. Once locked it will take some period for the GPS receiver to download enough data to be useful. How long depends on what data the GEO is sending at that moment. Often differential corrections can begin in 10 to 12 seconds from lock based on a need to download correction data for some of the SV's. At this point, even if the GEO is lost again, differential corrections will continue to be applied for about 2 minutes. The WAAS specification doesn't cover the case where a GEO satellite is drifting in and out of sync with the GPS receiver since, for airplane use, a clear view of the sky is assumed. Garmin seems extra conservative in dropping the differential corrections after only 2 minutes. While the GEO is in view the receiver will download corrections for additional SV's and the current ephemeris data for the GEO satellites themselves as well as correction data for the GEO satellites. Once ephemeris data is loaded the satellite bar will turn dark and it can also be used as part of the computed solution. Note that the GEO satellite can download correction data without, itself, being part of the GPS solution.

It seems that Garmin will favor differentially corrected satellites, when at least four exist, to the exclusion of regular satellites. If the four are in a poor geometric relationship the epe number, and possibly the accuracy of the solution, can be worse that it was with a regular solution. Luckily the GEO will generally download corrections for all of the satellites above a mask angle of 5 degrees so this anomaly is usually short lived. However, if the WAAS locks on a geo satellite that is outside the coverage area for that particular SV the results can be differential corrections based only on ephemeris and clocks which can result in poorer solutions than without WAAS. In this case the only recourse is to turn off WAAS.

If a GEO drops behind a hill the GPS receiver will lose its information just like any other GPS satellite. When the vehicle moves far enough the GEO may be seen again and will recover with its ephemeris data still current and re-enter the solution but it seems that ephemeris data on a GEO is shorter lived that on a normal GPS where it is good for hours. A loss of geo satellite dark bar status after only a few minutes of it being out of sight was observed. It could be that ephemeris data on a GEO is short lived since it tends to wobble a bit. And, thus is not truly in a circular orbit having much more eccentricity that a normal GPS satellite. Or, it could be that the GEO had been reset due to some ongoing test mode. More data will be required to determine which.

If you receive a satellite but do not have any ionospheric data for your area the Garmin receivers seem not to detect this condition. They will happily switch to differential correction mode and supply zero ionospheric corrections. This can cause the unit to show larger errors with WAAS enabled than without. They should apply the internal ionospheric correction algorithm in this case, but they don't seem to. So if you live in South America with no WAAS corrections be sure and turn WAAS off.

Chapter 17
Miscellaneous Functions

This is a collection of features found in Garmin receivers that are not especially connected with satellite navigation or unique to a small group of receivers. These functions may use satellite data and may be available on only one or a few of the models covered in this manual.

Celestial Data

Celestial Data includes Sun and Lunar data.

Sunrise and Sunset

This feature is found on all Garmin receivers. It computes the sunrise and sunset time for your location. On all units except the basic etrex it can also compute the sunrise and sunset times at other times and at other locations. The etrex shows this information as a selection field on the navigation page. The III family can show this information as a selection field on the position page. The III family can show this information for other locations or other times by using the planning feature in the route chapter. On other units the data is available as a selection option from the main menu.

From the main menu you select the Distance and Sun choice and the sunrise and sunset are displayed at the Destination location you choose. This page is intended to aid in planning a trip. You can use it to calculate distance for your trip as well as sunrise and sunset for any date you choose.

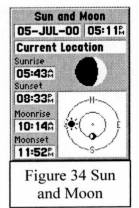

Figure 34 Sun and Moon

Sun and Moon

This feature is only found in the emap unit, the newest etrex units, the 76 family, and the GPS V. It can be reached from the main menu.

You can view the sunrise, sunset, moonrise, and moonset for your day and position or for any other day or position you care to set up. In addition there is a visual picture of exactly where the sun and moon are at the time you set. The phase of the moon is also shown as seen in the Northern Hemisphere. (Hopefully this will be corrected someday.) Any of the top three data fields shown on in the figure can be selected and changed.

- Date - The date field can be selected and changed to any date you wish.
- Day - The time field can be selected and changed to any time you wish.
- Current time - This is a local menu selection that will reset both the day and time.
- Location - Selecting this brings up a menu. You can select from current location, any map location, or from a location in the find menu.

You can use the diagram of sun and moon positions as a crude compass if you can see either the sun or the moon. This can be done by holding the unit in front of you pointing straight ahead. Now rotate your body until the sun or moon is in the same relative position as shown on the drawing if you imagine that you are in the center of the circles. When you have the sun or moon located in the correct position you will be facing north.

Special III family features

A battery monitor

A selection field is available on most screens to monitor the battery voltage and time. Time will be reset automatically when you change batteries. You can also set the type of battery you are using to provide a more accurate battery gauge setting.

Trip Planning

The route page contains a trip-planning feature. You can set fuel mileage and the planned date of the trip. You can build up a set of route points and use the selection field to view information about each one of them including sunrise and sunset at each point on the planned date. You can even plan your fuel stops.

Alarm Clock

While only the 12map contains an audible alarm any of the III family can set an alarm clock that will output a visual alarm at a preset time. Note that the unit must be on for this feature to work.

Special 12 family features

Surveying

While all Garmin receivers can use dgps, which can provide increase surveying accuracy with a suitable beacon source and any Garmin GPS can be used to provide for amateur surveying, the G-12 family has some special surveying features.

Area Calculation

The G-12 family can be used to perform area calculations. The idea is to traverse the area you are interested in and use the tracklog data to compute the total area of the track. This can be computed in acres and many other units and is available from the map page setup menu. There are a couple of things you should know.

- The full log is used so you should clear it before you start. If you don't close the loop then the program will work by computing the closed area obtained by connecting the last point in the log to the first point.
- The calculation uses signed numbers so if you walk the area and cross over your track at some point creating a figure 8 like picture then the program will compute the difference in area between two circles. However, if you are

257

careful and were to just touch the center but traverse in a single direction the area computed would be the total of both circles.

The etrex family

The etrex summit and etrex vista contains two features that are not in any other Garmin unit except the 76S. One is inclusion of a built-in electronic compass and the other is the inclusion of a built in electronic altimeter. In addition they share some nice features with other Garmin models such as celestial navigation and area calculation.

Electronic Compass

The summit and vista have a flux-gate compass built in that can be used to obtain your direction while you are stopped. Standard GPS units can only indicate the direction of your movement, which is derived from the standard GPS, fix information. A flux-gate compass uses magnetic field information just like a standard magnetic compass and can be used independent of any GPS data in the unit. However, the power of a combined unit is the ability to automatically switch the source of information as suitable for the circumstances.

The first step in using this compass is to calibrate it. This is done by selecting calibrate from the main menu and slowly turning around two times while holding the unit level. This may need to redone whenever the batteries are changed and may be required when the unit is located inside a vehicle. If you need to calibrate the unit for vehicle use you will need to drive the car in a circle twice. This step helps the unit account for any magnetic influences that may be present such as metal in the batteries.

Using the compass

The etrex summit has marks along the centerline to aid in using the compass to sight an object. One way to use the compass is to select 'sight and go' from the navigation local menu (press the enter key). Then hold the unit up to your eye level and keep it level in the palm of your hand. Sight the object you are interested in using the sighting aids and then carefully press the enter key without disturbing the position. The position will be locked in and you can lower the unit to read it. You can use the arrow keys to:

- Sight another position
- Project a waypoint from your present location. Select project and a waypoint entry screen will appear with the bearing already filled in and the distance field selected. You will need to enter the estimated distance. You can also change the waypoint name and icon. Select OK to save the waypoint or GOTO to save the waypoint and begin navigation toward the waypoint.
- Set Course to set a bearing you intend to follow. The compass will show a course deviation indicator by splitting and moving the center portion of the arrow to show if you drift off of the course you set. The data at the top of the screen will show exactly how far off you are. This is similar to the HSI screen show in the navigation chapter for the III pilot.

If you chose either of the navigation options shown above you can use the local menu to select Stop Navigation when you are finished following the pointer.

When following the compass you need to keep the unit level for best accuracy. If you get it too far from level it will give you a message. Once you exceed a preset threshold the unit will switch from using the magnetic compass to using the GPS as the compass display. This is usually 10 mph but it settable as a heading customization option from the main menu setup options. You can also set the switch back time so that if you temporarily reduce speed the unit won't be switching back and forth too much. You can force the unit to switch between magnetic and GPS compass by pressing and holding the page key.

The system setup entries permit turning the GPS off if you are only using the compass or turning the compass off if you are only using the GPS. Using only one or the other can save batteries. The magnetic variation setting can also be set. You can set the output to read true north, grid north, or magnetic north from system setup.

NMEA output

The Summit and Vista can output the electronic compass data via a standard NMEA sentence. This sentence looks like:

```
$HCHDG,A,B,C,D,E*chk

A = Magnetic sensor heading, degrees D.D
B = Magnetic deviation, degrees    not used
C = E / W                          not used
D = Magnetic variation, degrees
E = E / W    To get True add E variation to reading
chk = NMEA checksum
```

Altimeter features

The etrex summit and vista also have a built-in altimeter based on pressure changes. This is in addition to the altitude calculation that is performed as part of the GPS solution and can provide more accuracy. These units always use the altimeter for altitude readings and will always show a 3D solution even if you only have a 2D GPS solution. The altimeter altitude will be saved in waypoints and tracklogs. You can view the GPS altitude from a local menu item on the satellite page. The Altimeter altitude is present in the proprietary altitude sentence in NMEA mode while the GPS computed altitude is in the GGA sentence.

There is a special screen in the normal page rotation that is devoted to vertical navigation features. This screen will provide current elevation and an elevation profile similar the tracklog on the map page except that it is devoted to vertical movement. It will also display your vertical speed (ascent or descent). Selecting one of the two (on the vista) user selectable fields using the up/down arrow keys will permit choosing one entry from the following list: Local Pressure (renamed to Ambient pressure on new Vistas), Max Descent, Max Ascent, Average Descent, Average Ascent, Total Descent, Total Ascent, Min. Elevation, Max Elevation, (12 hour pressure trend for Summit), and for Vista vertical speed and normalized pressure (renamed to Barometric pressure on new Vistas), Glide Ratio, and Glide Ratio - Destination. The 12 hour pressure trend can be used to help predict the weather. Normalized pressure is adjusted to show what the pressure would be if the altitude was zero (Sea level).

Be careful on these units that you don't cover up the barometric pressure sensor which is located on the back of the unit in the curved area. Note the small hole in the center. If you cover this with your finger the pressure sensor will not be able to work correctly.

Note that if you change your altitude the pressure changes and this is the way the altimeter works to display the changes in altitude. Thus the normalized pressure will change also since there is not enough data to provide independent information on pressure changes and altitude changes, given that the GPS altitude is not heavily depended upon. You can use normalized pressure to predict weather changes if you remain in camp. If you use the GPS solution to maintain altitude calibration then the normalized pressure will shift less which will improve its ability to record weather related pressure changes.

A local menu on this page can be used to reset the elevation data, the max elevation, or select whether you wish the profile display to show time or distance, (or a pure 12 hour pressure display on Vista). You can also use the local menu to select zoom to zoom the screen in either direction, distance/time or elevation. A view points option permits using the up/down arrow keys to scroll through the data. Press enter to leave any of these modes.

Similar to the ability to show a saved tracklog visually on the map page, you can also show the tracklog vertical profile visually on the elevation page. This is an up/down arrow selection on the tracklog page.

The altimeter is generally expected to be automatically calibrated from the GPS altitude data, however you can calibrate it yourself if you wish. The main menu includes a calibrate option and you can select compass or altimeter. After selecting the altimeter you will be given up to three questions to calibrate the altimeter manually. These are: 'Correct the altitude?', 'Correct the pressure?', or 'Use the GPS altitude?'. Answer no until you get to the question you want and then answer yes to change the value. Note that the unit will display the current setting for each of these questions including GPS altitude for the last question so if, for some reason, you believe it is wrong you can answer this question with no as well and change nothing. (Note that this is a good way to view the GPS altitude, assuming you have a GPS fix of course.)

The automatic GPS calibration is not described in Garmin documentation, but here are some observations. Above about 15,000 feet it will correct the altimeter fairly rapidly once you have a lock fairly rapidly. On the ground it assumes you should be within 1000 feet of where the altimeter says you are and if the GPS altitude shows a bigger difference than that the unit will wait for you to correct it. Within 1000 feet it applies the GPS correction averaged over time and will correct half the error in about 22 minutes on an exponential curve that will eventually converge to the correct altitude. If you power it up with a different altitude after having been off for a while it will correct it in about 5 minutes. Of course if you don't have a 3D fix the altimeter altitude will be trusted to display a 3D solution. As a matter of fact these units never display the GPS computed solution and always indicate a 3D fix by using the built in altimeter. The tracklog also displays the altimeter altitude and this is what is used to display the vertical profile on the altimeter page.

The interface displays GPS altitude in the NMEA messages but the proprietary Garmin message displays the altimeter altitude. You cannot see the GPS computed altitude but it will be shown as a menu selection on the status screen or you go through the calibration process. You can cancel either activity, if you wish, after viewing the GPS computed altitude. (Note that early releases of the firmware used GPS altitude for waypoints but this is no longer the case.)

Other etrex features

The etrex Legend, Venture, and Vista contain a few more interesting features. For example they can be used to compute area and will product pseudo range signals similar to those already described in the 12 Family. It can also show Celestial Data as described above and some unique features described below.

Area Calculation

The area calculation works a little different from the version in the 12 series. For instance it takes advantage of the saved logs to do its calculation so this can be saved. In addition it automatically sets the units for the area based on whether you have set metric or English measurements, but you can change the units after the calculation is completed.

Generally the etrex expects you to walk (or drive) the exact route you intend to measure and will then compute the area live, at the time you finish, by connecting the last point to the first point to ensure the area is closed, and then performing the calculation. If you have crossed over your track during your travels it will not be correct. However it also computes areas for any of your saved logs even if they were not intended for this purpose. This area information should just be ignored unless you know the log meets criteria of no crossovers.

But what if you can't walk or drive the route because of terrain problems? The etrex can also calculate the area of a route.

1. Make waypoints of all 4 corners: a, b, c, and d. (or more if you wish)
2. Turn the set of waypoints into a route but you need not close the last point.
3. Select the route you just created and bring up the local menu. Select the route area to compute the area of the route you created.
4. Note that it will not be correct if the route crosses over itself.

Any existing route can also be used to compute an area. Select the route and use the local menu to select the calculate area command. Units can be changed by highlighting the current units and then selecting them to reveal a menu. Unless the route was specifically setup to compute an area it is likely to produce a meaningless result.

Calendar

The new etrex units have a calendar that can be used to display a date, jot down an appointment, or used as a method of reaching a particular period for celestial calculation or to find a best fishing time.

Hunt and Fish

The new etrex units can also calculate the best hunting and fishing times based on some undisclosed internal algorithm. It can be fun to calculate but don't yell at me if you don't catch a fish! While Garmin doesn't say how this is calculated you might find some answers if you study Solunar theory at: http://www.solunar.com/TheSolunarTheory.htm.

Calculator

The etrex line also includes a calculator on the accessories page. It works like most other calculators that you might own and include % and memory functions for the standard calculator and can be switched from the local menu into a full-fledged scientific calculator.

Jump Master

The Jump Master page, on the Accessories Menu for the Vista, is provided for Parachutists but perhaps glider pilots and balloonist could figure out some way to use it. It basically provides information to guide you to a target waypoint using the built in altimeter and compass on the Vista to help guide you down. It provides for 3 kinds of Jumps HAHO, HALO, and Static. These are defined to be High Altitude High Opening, High Altitude Low Opening, and Static jumps. High Altitude, Low Opening jumps in the military begin at about 25,000 feet and then free fall to about 3,500 feet when the parachute is opened. HAHO on the other hand begins at the same point but the parachute is opened immediately. A Static line Jump uses a line that automatically opens the chute about 12 feet into the jump. Another term used on the form is HARP, which mean High Altitude Release Point. The Desired Impact Point, DIP, is the place you would prefer to land, which needs to be an existing waypoint. The altitude you intend to use to open your parachute can also be entered. Altitude needs to be specified relative to the ground level altitude (AGL, altitude above ground level).

The winds with direction can be profiled on the Jumpmaster setup pages as well as constants such as K factor for the type of parachute, and desired accuracy for the DIP. The values entered are different for the different kinds of jumps. For example a static jump assumes a constant wind velocity.

76 family

The 76 family is a marine unit. It contains many of the special features of the etrex family of products and supports WAAS. A unique feature of this unit is the ability to support depth data as well as altitude data. The unit also supports more alarms than most other units described so far including anchor alarms and depth alarms. The depth feature is accomplished by supporting the NMEA DPT input sentence from a depth sounder. The depth data in integrated into the unit alarms, tracklog, and is stored in waypoint data. In addition NMEA output can be customized as needed to drive some autopilots.

The Map 76S has an electronic compass and altimeter. These have the same features as the etrex Vista unit described above. The altimeter in the 76S is a little more capable than the one in the Vista in that you can lock the altitude setting (for marine use) and cause the altimeter to behave like a traditional barometer. This is quite useful for weather predictions.

The altimeter/barometer in the 76S seems to be a better implementation than the one in the Vista. For example the normalized barometric pressure (which is simply called barometric pressure) is capable of adjusting itself correctly even when altitude changes. The unit will need to be on for about an hour before this seems to work properly, however once calibrated it can be a help in accurate weather prediction. It is also more sensitive than the one in the Summit and Vista indicating changes of only 5 feet.

Note that the compass needs to be held horizontally for operation while the Map 76S GPS wants to be held vertically for best performance. This can sometimes compromise the combined use of these two features.

GPS V

How about a game of breakout? This popular game is included in your GPS V in addition to a calendar, a hunt & fish calendar, Sun & Lunar data, and a regular general purpose calculator, which can all be reached from the accessories menu. These are described in the sections above.

In addition there is a Gas Mileage calculator that is unique to this unit. It automatically uses your trip distance although you can override this with your own distance data. To use the calculator bring it up from the accessories menu and then enter the amount of gas you purchased. It will compute and display your gas mileage. You can even press the menu button to reveal a choice to note this information on your calendar. As a convenience this page will also let you reset the trip meter so that it will be accurate for a future mileage check.

The breakout game uses the rocker pad to move the paddle at the bottom of the screen. The object is to remove all of the bricks.

Of course, the main unique feature of this product is the autorouting capability, which is covered in its own chapter.

Chapter 18
Working With Accessories

An external computer, laptop, palmtop, or desktop could be considered as an accessory for your GPS! This use is covered in another chapter devoted specifically to computer interfaces. A beacon receiver is also covered elsewhere. This section covers the minor accessories.

Cables

There are 3 types of cables for Garmin units. There are remote power cables, data cables, and combination cables that provide both services. It is also possible to obtain cables from a third party or get a connector and make your own. This is a unique third party shareware hardware product. Check with PFRANC. Most Garmin handhelds use a special circular connector with 4 pins. These are custom Garmin connectors but are available from third party sources. There are two versions of this connector. One with a small hole in the center and one without. The intent of the hole is to prevent a user from attempting to use a connector power cable designed for 12 Volt units on a unit that that can only accept up to 8 Volts. The 6 to 8 Volt units include the G-38, the G-40 and the G-12. These units have a plastic pin in the center of their cable receptacle to mate with the hole in the center of the cable connector. For data transfer either cable will work but for power the correct connector should be used. If you have a unit that wants 12 volts and have a connector with a hole in the center then it would be a good idea to fill it, just in case.

If you wish to wire your own cable or need to wire the far end of the cable you will need to know the wiring scheme. Look at the connector in the unit and find the locating guide. Proceeding clockwise the first pin is Data in, usually the white wire, the second is Ground, black, and is used as a return for both power and data. The third pin is data out, brown and the fourth is power, red. Garmin sells a data cable with the CPU end unwired so that you can use the connector of your choice for the computer connection. Check the interface chapter for a wiring diagram for round pin connectors.

The etrex and the emap need an entirely different connector that has all of the pins in a single line. The power is physically separated from the data connectors and the ground wire is on the opposite end. The data in connection is next to power while data out is next to ground. The power for these units is only 3 Volts so it also requires an external power adapter. Here is a diagram of the wiring for this connector. It shows the PFRANC version.

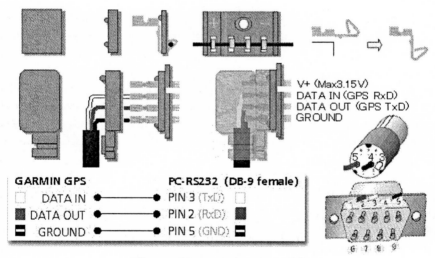

Figure 35 etrex /emap cable

External Power

There are many times when external power can be a useful accessory for your GPS. All of the Garmin handheld units have an external port available that can be used for power and for data, however some of the cables you purchase do not support both. The information in the cable discussion above and in the interface chapter provides the pin-outs for the power connection to make your own power connection if you wish. The ground connection is shared between power and data use.

Be careful! Wiring power to the unit can destroy the GPS if you input voltages that are not in the correct range for the unit. Also, if you get the polarity wrong you will destroy the unit. Power input is the positive (+) side of the power supply while ground is the negative (-) side.

When you buy a cable that includes a connector for a cigarette lighter you might assume the unit itself uses 12 Volts. This is often not the case as there are plenty of 12 Volt adapters that will fit inside the cigarette lighter housing so do not cut off a cable and assume you can just wire it direct to 12 Volts. Check the Technical Specs chapter for details on the power requirements for various models or look in the manual that came with the unit. However, often you may also want to run the unit from an AC outlet. If you already have a 12 volts adapter cable then you should buy an AC to 12-Volt power supply with a cigarette lighter receptacle and then plug in the 12-Volt adapter you already have. Radio Shack sells these kinds of adapters.

External power is isolated from the internal batteries such that you cannot charge the batteries from the external power source. This is by design since Garmin cannot know what kind of batteries may be present in the unit and trying to charge the wrong kind can damage the unit. However some units, notably the G-12, are not isolated to the extent that you cannot continue to use the batteries even when external power is applied. The isolation is done with a diode, so on these units you need to supply at least 6.7 Volts to avoid partial discharge of the internal batteries. Garmin's own external power supply provides close to 8 Volts to these units. If you really want to charge a set of batteries, which they are inside the unit, then you may want to snake a couple of wires inside the battery compartment itself and connect them to the unit.

Data Port Accessories

One very useful hardware attachment for NMEA use is a datalogger. The idea of a data logger is to collect data in real time from the GPS and then save it for later processing. One such datalogger is available from http://homepages.tig.com.au/~robk/datalogger.html. It will log up to 270,000 points from the NMEA $GPRMC sentence as fast as the sentence is generated or every 10 seconds as chosen by the user. There are other data loggers available as well. Other more specialized units are used by glider pilots to verify their flights in contests. One such unit is made by EW Avionics. Another hardware attachment is ERIC. This is a hardware attachment that reads route (RMB) sentences available through the NMEA interface and report your progress verbally. It hooks directly to the data port and can share the port with other devices. You load a route into the GPS and activate it. ERIC will then warn you about the waypoints in the route as you near them. This is especially useful if your GPS doesn't have audible alarms. (The web page for this site seems to have vanished.)

There are plenty of other devices that work with NMEA output from a Garmin GPS. These include flight computers, boat autopilots, depth finders, PDA's, and more. The GPS can be used to steer a boat from the NMEA data when set up in Navigation mode.

Some digital cameras have been designed to accept GPS data for imprinting as part of the image. Generally most of these cannot accept NMEA data but require a simple text output. Some Garmin receivers have the ability to output this simple text format. These include the III+, the 12Map and the etrex units. The newest versions can now accept NMEA data as well.

Some Garmin units can accept data input from NMEA compatible devices as well. For example the 76 marine units will accept input from a depth finder unit and will then display and record depth data in the unit. Internal alarms can also be set based on this data.

Antennas

Many Garmin GPS receivers can accept a remote antenna. Generally these are active antennas meaning that they have pre-amplifiers inside and Garmin supplies power through the connector to run the antenna. If you wish to hook an unpowered antenna to the external antenna connector you will need to block the dc power using a capacitor to avoid damaging the unit. Garmin supports two kinds of remote antennas. One uses a very small mcx connector and when plugged into the unit it removes power to the internal patch antenna. To do this Garmin has a circuit inside that sense the power drawn by the external unit and switches the antenna source. The external antenna needs to draw at least 5 ma for this switch to take place. The second kind of remote antenna is used with the Garmin II and III families. These units already have an external antenna attached to the backside. You simply remove the supplied antenna and attach a replacement. The connector in this case is a BNC connector. In this configuration there is no need to sense and switch the antenna since only one is plugged in at a time.

Since the helix antenna is simply plugged into the back of the unit, it can also serve as a remote antenna. In this case you only need a cable to remote mount the antenna. Garmin even sells a short cable with a suction cup mount at one end that can be used to place the supplied antenna a short distance from the unit. You could also easily make your own. Note that the helix antenna is not amplified and thus cable losses will limit the maximum length of this cable to about 6 feet.

Most Garmin receivers that support remote antennas supply 5VDC in the cable to operate the antenna. The newest units, the 12Map, the e-map, and the 76 family, only supply about 2.5VDC to the antenna connector. This must be considered when choosing an antenna. Since the remote antenna uses power you can expect a shorter battery life whenever using a remote antenna.

The following handhelds do not have a provision for a remote antenna: The G-38, the G-12, the etrex, and the just released 72. For these units or for any unit where you don't want to take advantage of the built in antenna extension you can purchase a re-radiating antenna. A re-radiating antenna is really two antennas. One amplified receiving antenna collects the GPS signals and sends it to a second antenna that re-transmits the signals. The GPS receiver is placed close to this second antenna to pick up the signals using its own built in antenna. An external power source is needed to power the re-radiating antenna independent from the GPS receiver. One such antenna is available from <u>Tri-M Systems</u>.

One question that usually arises is: "Do I need an external antenna?" You can answer this question yourself by observing the behavior of your unit on the satellite status screen. The screen displays all of the satellites your unit could be expected to observe at any point in time. If you are receiving all of them with reasonable strength or are missing only those that are obviously behind buildings or hills then you will not benefit from an external antenna. However, if reception is spotty or weak and you observe this behavior while moving under tree cover or other light cover conditions then an external antenna may help your reception considerably.

Cases/brackets

Garmin sells cases and mounting brackets for car, boat, and bicycle use. In addition you can get these accessories from independent sources. In particular cell phone accessories will often work just fine with a GPS and marine supply stores sell many waterproof cases and brackets that can provide for unusual mounting conditions. A zip lock bag can provide some waterproof insurance for your unit. For the emap there is also a neoprene glove like protective case that covers most of the unit.

The bicycle brackets from Garmin come in two pieces. One is attached to the unit and the second is attached to the bike. The bike piece is a small ring that is intended to be left permanently on the handlebars. The piece that attaches to the GPS varies from one unit to another. The most interesting is the etrex which replaces the battery cover with a different cover that contains a plastic clip that is designed to slide into a slot on the bicycle ring piece. The cover also has a rubber casket that improves the water resistance of the battery compartment, which may make it useful even if you don't want to use it with a bike.

The G-12 family, the 12 map and the older multiplex units have a similar case use a plastic clip that permanently replaces the rubber backing on these units. This clip mates with a slot in the ring like the etrex unit. Since this clip is permanent it does make the unit slightly more awkward to lay on its back and prevents its use in one of the car/boat mounts available. The emap uses a similar mounting idea that screws into the back of the unit.

The III series uses a more complicated clamp that clips onto edges of the unit and can be held in place with a screw. This clamp, in turn, slides into the slot in the ring. Some have reported that the clamp can break under vibration so this mounting should include a safety hook up.

The emap has a special yellow slip-on neoprene sleeve that can improve its ruggedness when used in outdoor environments by providing some shock protection and some additional protection against water.

Memory Cartridge

Of the handhelds described in this manual only the Emap has a provision for add-on memory cartridges. You can have a maximum of one cartridge installed at any one time. The emap requires an optional cartridge to permit any add-on maps and this cartridge holds these maps and the poi database associated with these maps. There is a large selection of maps available and they are described in the database chapter.

Memory cartridges are available with maps already installed in 4 Meg, 8 Meg, and 16 Meg sizes. You can also purchase blank cartridges in 8, 16, 32, 64, and 128 Meg sizes. Any of the cartridges can be programmed from the available cdroms. To install a cartridge you will need to open the battery compartment and remove the batteries. You will then see a small opening in the upper right corner of the compartment. Insert the cartridge into this opening with the label facing up. To remove you need to use a fingernail to pull out the extraction handle and then you can remove the cartridge.

The cartridge extractor is attached with a small screw. Be sure that this screw is tight and flush with the side of the unit or you may have problems attempting to extract the cartridge.

Garmin has also released a USB programming module for the cartridges. In the larger sizes the ability to download maps at USB speeds can significantly speed up the process. Once the cartridge is programmed it can be transferred to the emap unit.

Chapter 19
Garmin Technical Information

This chapter provides technical specifications for each of the Garmin units covered in this manual as well as troubleshooting information.

Technical Specifications

This information is presented in tables containing a comparison of the various products in the Garmin handheld GPS line. Instead of building a table entry for each product they have been grouped by family. The older multiplex units are represented by the G-38/G-40/G-45XL and the 12 channel parallel units are represented by the G-12/G-12XL/G48. The original mapping products are shown in the table entry for G-III/GIII+. There are special columns for the emap, etrex lines and 76 family. The etrex has two columns due to the split in the family coverage based on the click stick hardware. Note that the European version of the basic etrex is really the equivalent of the Camo model in the USA. There is also some support for an electronic compass and an altimeter with vertical profiling. These are unique features in a few Garmin units. There is no specific coverage in the table for the G-II, the G-II+, the G-III Pilot, the G-12CX, the 12MAP, or the GPS-72. Instead these units are represented as note entries at the bottom of the table. The specifications for other Garmin units may be able to be implied from the table but no guarantee can be made for their applicability.

General Notes for table.
- Y means feature is present, N means not present.
- A model number means feature is only on this unit.
- N/A means this feature is not applicable or not needed.
- There are some features of Garmin products that are not addressed in this table.
- See notes at end of table for other models.

Table 5 – Technical Specifications

Feature	38, 40, 45XL	12, 12XL, 48	III, III+	V	Emap	Etrex, Camo, Summit	Venture, Legend, Vista	76, Map76, Map76S	Notes
Waypoints	250	500	500	500	500	500	500	500, 76S has 1000	Note 1
Waypoint names	6	6	6	15	10	6	10	10	Characters
Waypoint comments	16	16	16	50	N	N	N	N	Characters
Waypoint Altitude	N	N	N	Y	Y	Y	Y	Y	.
Waypoint icons	N	16	74	78	75	29	77	75	
Nearest Waypoint	9 < 160 km	9 < 160 km	9 < 320 km	10	15 < 130 km	9 < 320 km	15 < 130 km	10	Note 12
Project Waypoint	Y	Y	Y	Y	N	Y	Y	Y	
Favorite Waypoints	N	N	N	N	N	N	Y	N	
Previous Points	N	N	N	Y	N	N	N	N	list of recently used points
MOB	Y	Y	Y	?	N	N	N	Y	Man Over Board
Routes	20	20	20	20	50	1, 20 on summit	20	50	Reversible
Waypoints / route	30	30	30	50	50	50	50	50	
Route numbers	Y	Y	N	N	N	N	N	N	

Feature	38, 40, 45X L	12, 12XL, 48	III, III+	V	Emap	Etrex, Camo, Summit	Venture, Legend, Vista	76, Map76, Map76S	Notes
Route names	16	16	16	> 32	13	13	13	21	characters
Trip Planner	N	N	Y	?	N	N	N	Y	
Auto Routing	N	N	N	Y	N	N	N	N	
Popup Directions	N	N	N	Y	N	N	N	N	
2D mode	Y	Y	Y	Y	Y	Y (not summit)	Y (not vista)	Y	Only if 3D is unavailable
Overde-termined Solution	N	Y	Y	Y	Y	Y	Y	Y	can improve accuracy
Horizontal Accuracy	15	15	15	15	15	15	15	15	meters.
DGPS ready	Y	Y	Y	Y	Y	Y	Y	Y	
Remote Control of beacon receiver	Y	Y	Y	Y A	Y	Y	Y	Y	A means that automatic mode is available
DGPS Accuracy	5-10	1-5	1-5	3-5	1-5	1-5	1-5	3-5	meters
WAAS	N	N	N	Y	N	N	Y	Y	accuracy < 3 Meters
Speed Accuracy	0.1	0.1	0.1	0.1	0.1	0.1	0.1	0.1	knots
Warm Start	20	15	15	15	15	15	15	15	seconds
Cold Start	120	45	45	45	45	45	45	45	
EZ-Init	S	S	M	M	M	No	M except venture = C	M except 76 = C	S = state or country, M = Maps, C = city.

Feature	38, 40, 45X L	12, 12XL, 48	III, III+	V	Emap	Etrex, Camo, Summit	Venture, Legend, Vista	76, Map76, Map76 S	Notes
Autolocate	15	5	5	2	5	5	5	5	minutes
Tracked Satellites	8	12	12	12	12	12	12	12	upto
Water-proof	NS	IPX7, G-48 IPX4	IPX7	IPX7	IPX2	IPX7	IPX7	IPX7 floats	Note 17, NS = nitrogen sealed
Dynamics	3G	6G	6G	6G	6G	6G	6G	6G	
Units	NSM	NSM	NSM	NSM	NSM	NSM	NSMY	NSM	N=Nautical, S=Statute, M=Metric, Y=Yards
City Database	N	12XL 48 12CX	Y	Y	Y	N	Y	Y world-wide	
Nearest Cities	N	N	50	Y	50	N	50	50	
Navaid Database	N	48	III+*, 12Map*	Y*	Y*	N	Y	Y	* using optional maps
POI Database	N	N	N	Y*	Y*	N	Y	Y	* using optional maps
Weight	9.5 oz	9.5 oz	9.0 oz	9.0 oz	6.7 oz	5.3 oz	5.3 oz	7.5 oz	with batteries
Case	wel-ded	wel-ded	Gas-ket	Gas-ket	welded	Welded	welded	Gasket	
Temp Range	5F to 158F	5F to 158F	5F to 158F	5F to 158F	5F to 158F	5F to 158F	5F to 158F	5F to 158F	Note 1, operating
Screens	5 or 6 from 7 choices	5 or 6 from 7 choices	6 or 7	4 from 5 choices	4	4 from 5 choices +1 on summit	5 from 7 choices +1 on vista	5 +1 on Map76S	Note 10,11, 13
Screen Size	2.2x 1.5	2.2x 1.5	2.2x 1.5	2.2x 1.5	2.2x 1.65	2.2x 1.2	2.2x 1.2	2.2x 1.6	inches
Screen Resolution	64x 100	64x 100	100x 160	160x 256	120x 160	64x 128	160x 288	180x 240	.
Screen Density	1,939	1,939	4,848	12,412	5,289	3,103	17,454	12,273	dots/sq. inch
Grey Scale	N	N	Y	Y	Y	Y	Y	Y	Note 1,4

Feature	38, 40, 45XL	12, 12XL, 48	III, III+	V	Emap	Etrex, Camo, Summit	Venture, Legend, Vista	76, Map76, Map76S	Notes
Battery Life	17	30	36	25	14	22, 16 on summit w/ compass	20 venture, 18 legend, 12 vista w/ compass	16 hours, 12 Map76S w/ compass	Note 1, normal mode is approx. 2/3 battery save
Batteries	4	4	4	4	2	2	2	2	AA cells
Backup battery	Y	Y	Y	N/A	N/A	N/A	N/A	N/A	3 months - rechargeable lithium with a ten year life.
External power (low voltage)	38 40 (5-8V)	12 (5-8V)	N	N	3-3.3 V	3-3.3 V	3-3.3 V	N	All need an external voltage regulator
External power 10-32V	45XL II	12XL 48 12CX II+ 12MAP	Y	Y	N	N	N	8-35V	All use Garmin Round connector
Dimension	14.6 x 5.1 x 3.4	14.6 x 5.1 x 3.4	12.7 x 5.9 x 4.1	12.7 x 5.9 x 4.1	14 x 6 x 2	11.2 x 5.1 x 3.0	11.2 x 5.1 x 3.0	15.7 x 6.9 x 3.4	cm
Number of keys	10	10	12	12	12	5	10	12	Note 1, rocker = 4 keys
Sun/Moon positions	N	N	N	N	Y	N	Y	Y	North finder
Map Screen Orien-tation	N, T	N, T, C	N, T, C	N, T	N, T	N, T	N, T	N, T, C	N=North up T=track up C=course up
Main Tracklog	768	1024	1900	3000	2048	1536, 3000 on summit	2048, 3000 on vista	2048, 5000 on Map76S	
Saved Tracklogs	N	N	10	10	10	10	10	10	Note 2

Feature	38, 40, 45XL	12, 12XL, 48	III, III+	V	Emap	Etrex, Camo, Summit	Venture, Legend, Vista	76, Map76, Map76S	Notes
Saved log	N/A	N/A	250	250	250	125, 500 on summit	250	250	
Disconti-nuous Tracklogs	Y	Y	Y	Y	Y	Y	Y	Y	Saved logs are continuous
Tracklog collection	A, T	A, T	A, T, D	A, T, D	A	A	A, T, D	A, T, D	A=Automatic T=Time D=Distance
Tracklog modes	O, W	O, W, F	O, W, F	O, W	W	W	O, W, F	O, W, F	O=Off W=Wrap F=Fill w/alarm
Altitude in tracklog	N	N	N	Y	Y	Y	Y	Y + depth	.
Area calculation from track log	N	Y	N	Y	N	N	Y	N	.
Backtrack	via route	via route	via route	saved track-log	saved track-log	saved tracklog	saved tracklog	saved tracklog	.
Dead Reckoning	30 secs.	30 secs.	30 secs.	30 secs.	30 secs.	30 secs.	30 secs.	30 secs.	.
Project Track	N	N	N	N	N	N	N	N	
Pan & Zoom Map screen	Both	Both	Both - zoom keys	Both - zoom keys	Both - zoom keys	Zoom only	Both - zoom keys	Both - zoom keys	Note 1,4
Fish/Hunt calculator	N	N	N	Y	N	Camo	Y	Y	When to fish/hunt, Note 14
Sunrise/ sunset	Y	Y	Y	Y	Y	Y	Y	Y	.
Moonrise, moonset, phase	N	N	N	Y	Y	Camo	Y	Y	Note 14
Calculator	N	N	N	Y	N	N	Y	Y	standard and scientific

Feature	38, 40, 45X L	12, 12XL, 48	III, III+	V	Emap	Etrex, Camo, Summit	Venture, Legend, Vista	76, Map76, Map76 S	Notes
Position averaging	N	Y	Y	Y	Y	N	N	Y	horizontal only except emap, 76
Daylight Savings Time	M	M	M	A	A	A	A	A	M=Manual units you must change time zone, A=Automatic can be disabled
Quadrifilar Antenna	45 XL	48	Y	Y	N	N	N	Y	Note 5,6, removeable except 76 series
Patch Antenna	38 40	12 12XL 12MAP	N	N	Y	Y	Y	N	
Optional External Antenna	40 45 XL	12XL 48	Y 2.5 V on 12Map	Y 2.5 V	Y 2.5 V	N	N	Y 2.5 V	5 V power except as noted. also Note 1,4,5,6
Backlit lamp levels	1	3	3	2	1	1	1	1	
Trip Odometer	Y	Y	Y	Y	Y	Y	Y	Y	.
A Second Odometer	N	N	Y	Y	Y	N	Y	Y	.
Max Speed	N	Y	Y	Y	Y	Y	Y	Y	.
Speed Limit	99	999	999	999	999	999	999	999	Knots
Average Speed (moving)	N	Y	Y	Y	Y	Y	Y	Y	Note 3
Average Speed (overall)	N	N	Y	Y	Y	Y	Y	Y	Note 3

Feature	38, 40, 45X L	12, 12XL, 48	III, III+	V	Emap	Etrex, Camo, Summit	Venture, Legend, Vista	76, Map76, Map76 S	Notes
Tunable Speed Filtering	N	G-48	N	N	N	N	N	Y	.
Trip Time	N	Y	Y	Y	Y	Y	Y	Y	.
Elasped Time	N	Y	Y	Y (trip)	Y (trip)	N	Y (trip)	Y (trip)	.
Direction	M, T, Grid Deg-rees	M, T, G, U, deg-rees / mils	M, True deg-rees	M, True deg-rees	M, T, G, deg-rees	M, T, G Degrees	M, T, G, U, degrees	M, T, G, U, degrees	M=Magneti c T=True G=Grid U=User
Built in Compass	N	N	N	N	N	Summit	Vista	Map76S	Can also be derived by GPS when moving
Compass Accuracy	N/A	N/A	N/A	N/A	N/A	+/- 5	+/- 5	N/A	1 degree resolution
Altimeter	N	N	N	N	N	Summit	Vista	Map76S	Can also be calculated by GPS
Altimeter Accuracy	N/A	N/A	N/A	N/A	N/A	10	10	N/A	1 foot resolution
Download routes	Y	Y	Y	Y	Y	Y	Y	Y	Upload also
Download Waypoints	Y	Y	Y	Y	Y	Y	Y	Y	Upload also
Download tracklog	Y	Y	Y	Y	Y	Y	Y	Y	Upload also
Audible Alarms	45 XL	12XL 48 12CX 12 MAP	N	Y	Y	N	N	Y	All have visual alarms
Anchor Alarm	N	48	N	Y	N	N	N	Y	
Proximity Alarms	9 45 XL	9	N	10	N	N	N	10	

Feature	38, 40, 45XL	12, 12XL, 48	III, III+	V	Emap	Etrex, Camo, Summit	Venture, Legend, Vista	76, Map76, Map76S	Notes
Arrival Alarms	1 min warning or distance	1 min warning or distance	1 min warning or distance	15 sec and up scaled to speed	15 sec and up scaled to speed	15 sec	15 sec	15 sec and up scaled to speed or distance or time	Automatic rollover to next route leg
Simulation Mode	All	All	except setting location	All	All	speed is fixed.	speed is fixed	All	All units support this
Language Support	1	9	1	6	5	1, 12 summit, 13 camo	1	7	note 14, 15
Base Map	N	N	Y	Y	Y	N	Legend, Vista	76Map, Map76S	.
Uploadable Maps	N	N	III+, 12 Map	Y	Y	N	Legend, Vista	76Map, Map76S	Emap needs a cartridge for maps
Data Memory	N O N E	Unknown	1.4 Meg	19 Meg	Optional	None	1M Venture, 8M Legend, 24M Vista	1M 76, 8M 76Map, 24M Map76S	emap optional cartridges 8, 16, 32, 64, 128 Meg
Tide Data	N	N	N	N	N	N	N	Y	Americas only
Display Customization	few	Some	Extensive	Extensive	None	None	Good	Extensive	Note 6
Display Orientation	V	V	V / H	V / H	V	V	V	V	V = Vertical, H = Horizontal Note 4,5,6
Interface outputs	G, N	G, N	G, N, T see note	G, N, T	G, N, T	G, N, T	G, N, T	G, N	G=Garmin N=NMEA T=Text output III+, 12map

Feature	38, 40, 45X L	12, 12XL, 48	III, III+	V	Emap	Etrex, Camo, Summit	Venture, Legend, Vista	76, Map76, Map76 S	Notes
Interface inputs	G, N, R	G, N, R	G, N, R	G, N, R	G, N, R	G, R	G, N, R	G, N, R	G=Garmin N=NMEA R=RTCM
Datums	104	107	105	> 100	107	> 100	> 100	> 100	All have 1 user defined

Notes:

1. 12CX is a 12XL except it has 2048 tracklog, 3 color display, 1000 waypoints, waypoint maintenance improvements, 35 hour battery life w/battery save, dedicated zoom keys, temperature 5 - 131 F
2. III Pilot is like a III except - no saved logs, no MOB, has Jeppesen database support, HSI compass display, vertical navigation support
3. III and III+ have configurable average speed.
4. 12MAP - A 12CX hardware and case design with audible alarms. Software features and display are the same as a III+ except no display rotation.
5. II is the same specs as other multiplex units except it in a III case design with rotatable display
6. II+ is like a III except no maps and has lower resolution display and much less customization.
7. keypad if used counts as 4 keys but may be used for 8 positions on units without separate keys.
8. Information for final destination available on G12 family and G45 family only on active route screen. Not on a navigation screen. Etrex and emap report this on backtrack routes. Emap reports this on route page.
9. Etrex, emap reports on the active route screen, but active backtracks show ETE and Dist on main screen. Note that emap, etrex calculates this differently from other units.
10. Etrex summit, vista, 76S contains another screen for vertical profiling. Can be customized to display barometric profiling.
11. Etrex and Emap display some live data on menu screens. Both will display active route data and the emap has a trip

screen that displays average speed, max speed, and odometer read outs. The setup screen is the only place on the emap to display time in seconds.

12. Emap and new etrexes compute nearest from cursor position - the others from present position.

13. GPS V - satellite screen, map, compass (position), trip guidance.

14. The European basic etrex is like the USA Camo except for color.

15. Based on the US version, you can buy some units in other languages.

16. The GPS 72 is like the 76 except there is no provision for an external antenna and the screen resolution is 120x160.

17. IPX7, waterproof in 1 meter of water for 30 minutes, IPX2 splash proof, battery compartments are generally not waterproof.

All Garmin units support English/Metric/Nautical units for speed, distance, and altitude.

Table 6 – Grids and Navigation Data Supported

Feature	38, 40, 45XL	12, 12XL, 48	III, III+	V	Emap	Etrex, Camo, Summit	Venture Legend Vista	Notes
Grids								
DDD.ddddd	Y	Y	Y	Y	Y	Y	Y	
DDD/MM.mmm	Y	Y	Y	Y	Y	Y	Y	
DDD/MM/SS.s	Y	Y	Y	Y	Y	Y	Y	
TD	N	Y	Y	Y	N	N	Y	Loran
UTM/UPS	Y	Y	Y	Y	Y	Y	Y	
MGRS	N	Y	Y	Y	Y	Y	Y	Military
OSGB	Y	Y	Y	Y	Y	Y	Y	Great Britain
Irish	Y	Y	Y	Y	Y	Y	Y	
Swiss	Y	Y	Y	Y	Y	Y	Y	
Swedish	Y	Y	Y	Y	Y	Y	Y	
German	Y	Y	Y	Y	Y	Y	Y	
French	N	N	N	N	N	N	N	
Indo So LCO	Y	Y	Y	N	N	N	N	
India	N	Y	Y	Y	N	N	Y	O, IA, IB, IIA, IIB, IIIA, IIIB, IVA, IVB
Maidenhead	N	Y	III+	Y	Y	Y	Y	
New Zealand	N	Y	N	Y	Y	Y	Y	

Feature	38, 40, 45XL	12, 12XL, 48	III, III+	V	Emap	Etrex, Camo, Summit	Venture Legend Vista	Notes
Taiwan	Y	Y	Y	Y	Y	Y	Y	
W. Malayan RSO	N	Y	N	Y	Y	Y	Y	
Dutch Grid	N	N	N	Y	Y	Y	Y	
Qatar Grid	N	N	N	Y	Y	Y	Y	
Finish KKJ27 grid	N	N	N	Y	Y	Y	Y	
User Grid	N	Y	Y	Y	Y	Y	Y	
Navigation Data								
Location	Y	Y	Y	Y	Y	Y	Y	lat/lon or grid
Dual grids	N	N	N	N	N	Y	N	Lat/Lon must be one
Altitude	Y	Y	Y	Y	Y	Y	Y	
Speed	Y	Y	Y	Y	Y	Y	Y	
Track	Y	Y	Y	Y	N	Y	Y	aka Course
Bearing	Y	Y	Y	Y	Y	Y	Y	
ETE	Y	Y	Y	Y	Y	Y	Y	time to go
ETA	Y	Y	Y	Y	Y	N	Y	
XTE	Y	Y	Y	Y	N	N	Y	Course error
CTS	Y	Y	Y	Y	N	N	Y	To Course
VMG	Y	Y	Y	Y	N	N	Y	
TRN	Y	Y	Y	N	N	N	Y	

Feature	38, 40, 45XL	12, 12XL, 48	III, III+	V	Emap	Etrex, Camo, Summit	Venture Legend Vista	Notes
ETAD	N	N	Y	Y	Y	N	Y	Note 8,9 ETA at Destina-tion
ETED	N	N	N	Y	N	N	Y	ETE at Destinati on
Dist to Next	Y	Y	Y	Y	Y	Y	Y	.
Dist to Destina-tion	N	N	Y	Y	Y	N	Y	Note 8,9
VS	N	N	N	Y	N	Summit	Vista	Vertical Speed

Error Recovery

From time to time you may get an error when using your GPS receiver. Most of the time it will give you a message, which may not be obvious to a new user. This chapter deals with the messages and other error symptoms. Not every receiver will produce all of the messages listed below.

Messages

The messages below are listed in alphabetical order. Some specialized III Pilot messages are not listed.

1. Active WPT Can't be Deleted - The waypoint you are trying to delete is the destination of a goto or being used in the current leg of a route. You must cancel the goto or stop the route to delete this waypoint.

288

2. Accuracy has been Degraded - The satellite geometry or data quality has resulted in an epe of more that 500 meters. Don't trust the GPS solution.

3. Already Exists - The name you have chosen is already in use. Pick another name.

4. Anchor Alarm - Boat has drifted outside the area defined in the anchor alarm setting.

5. Approaching ... - You are one minute away from a turn or the listed destination.

6. Approaching Turn - You are about 15 seconds from the next turn.

7. Arrival at ... - You are one minute away form the listed destination.

8. Arriving at Destination - You are within 15 seconds of your destination. If you power off the unit now the route will be terminated.

9. Base Map Failed - There is an error in the base map database. The unit must be returned to Garmin.

10. Batteries Low - see Battery Power is low

11. Battery Power is Low - Batteries are about to run out. They should be replaced at your earliest convenience. Depending on the type of batteries and unit you use you will have anywhere from a minute to about 30 minutes to replace the batteries.

12. CDI Alarm - You are left or right of course more that the limits set on the CDI alarm menu.

13. Data Transfer Complete - see transfer complete

14. Database Error - There is an error in the unit's database. Generally if this repeats you will need to return the unit to Garmin. You may be able to reset the unit or reload the firmware.

15. Degraded Accuracy - see Accuracy has been Degraded

16. GPS Turned Off (for use indoors). The unit is operating in simulation mode.

17. Leg not smoothed - The upcoming turn is too sharp to perform a smooth waypoint transition.

18. Lost Satellite Reception - see Poor GPS coverage

19. Memory Battery Power Low - The backup battery is low. This battery is recharged by the normal batteries. You need to put a set of new batteries in the unit and wait 24

hours to see if this message goes away. If it does not this means your backup battery is bad and data will be lost if you remove the main batteries. You will need to return the unit to Garmin for a new backup battery.

20. Need Altitude - Unit cannot obtain a fix with the current assumption about the altitude. You will need to give it a new altitude that is closer to correct.

21. Need to Select Init Method - Unit cannot locate satellites to compute a solution. You will need to select the initialization method. You have probably traveled more than 300 miles since it was last turned on or the unit has been off for more than 3 months. Perhaps you started with the unit in a house and can now chose Continue to acquire.

22. No DGPS Position - see No Differential GPS Position

23. No Differential GPS Position - You are trying to use a beacon receiver as set up in the interface menu but the unit is not able to communicate with the receiver, or there is not enough data to compute a DGPS position.

24. No RTCM Input - The Beacon receiver is not hooked up correctly or the baud rate is set wrong.

25. None Found - The unit cannot find an address, intersection, or point of interest that you were searching for. Modify the search argument as needed and try again.

26. Off Course Alarm - see CDI alarm.

27. Poor GPS Coverage - The unit cannot track the minimum of three satellites required and has not been able to do so for the last thirty seconds. You will need to find a place with a clearer sky view to recover from this error. Acknowledging the error does not change the status. Note that there will be a break in the tracklog as a result of this loss of lock. Any dead reckoning performed by the unit just prior to the loss will be removed from the log.

28. Power Down and Re-init - The unit is not able to calculate a position solution from the available data. Do what it says. You may need to try a new location.

29. PROX alarm - you have entered the alarm circle defined for the proximity waypoint.

30. Proximity Overlapped - Two proximity alarm circles overlap. This could cause problems monitoring the distance from each.

31. Proximity Wpt can't be Deleted - The waypoint you are trying to delete is listed on the proximity alarms page. You must remove it there first.

32. RAM Failed - The internal RAM memory has failed. If this persists you will need to return the unit to Garmin.

33. Read Only Mem has Failed - see ROM failed

34. Received an Invalid WPT - A waypoint was received during upload with an invalid identifier

35. Receiver has Failed - A hardware failure as been detected. If this message persists the unit will need to be returned to Garmin

36. ROM Failed - The permanent software memory has failed and the unit is not operable. You will need too return the unit to Garmin.

37. Route is Full - Attempted to upload a route with to many waypoints.

38. Route is not Empty - Attempted to copy a route into one that was already in use.

39. Route Memory Full - There is no more memory available for routes. You will have to delete some routes in order to add more.

40. Route Waypoint Can't be Deleted - Waypoint is part of a route. It must be deleted from the route before it can be deleted.

41. Route Waypoint was Deleted - A route waypoint entered does not exist in the database and was, therefore, deleted from the route.

42. Route Truncated - You were uploading a route that has too many points in it.

43. RTCM Input has Failed - DGPS data has been lost. You are no longer receiving the beacon signal. Check the cable.

44. Searching the Sky - The unit is searching the sky for almanac data or it is in Autolocate mode.

45. Steep Turn Ahead - The message appears when the course change is too abrupt for the speed you are traveling. Generally this message will start to appear above 75 mph and is intended for planes.

46. Stored Data was Lost - All waypoints, routes, time and almanac data has been lost. Could be a backup battery failure or a full reset.

47. Timer Expired - The count down timer has reached zero.

48. Track Log Already Exists - The name selected is the same as an existing tracklog. You will need to pick a different name.

49. Track Memory Full - All of the saved track locations are used.

50. Track Truncated - You were uploading a tracklog that is too long for the track memory. You may be able to clear the track log and try again unless the uploaded log is longer than the full length of track log.

51. Transfer Complete - This message means the upload or download of data was successful.

52. Transfer has been Completed - see Transfer Complete

53. Trouble Tracking Satellites, Are you indoors? - Message means that the unit cannot get a fix. You can answer Yes, which means you will be switched to simulation mode, or No which will cause a series of questions to determine whether or not to enter auto-locate mode.

54. Using Simulator for Faster Data Transfer - The PC software is selecting simulation mode to enable a higher transfer rate.

55. Waypoint Already Exists - You tried to make a new waypoint with the same name as an existing waypoint. Name the new waypoint something else.

56. Waypoint Memory Full - The full number of waypoints is used. The number of waypoints you can have is dependent on the model. Some have 250, some 500, and some 1000. You will have to remove some waypoints to make room for any new ones.

57. Weak Signals – see poor GPS Coverage

58. WPT Memory is Full - see Waypoint Memory Full.

Trouble Shooting Problems

Here is a table that covers troubleshooting methods for some of the problems that might be encountered when using Garmin units.

Table 7 Failure Symptoms

Symptom	Trouble Shooting tips
Fails to highlight any satellites	If you have moved more than 300 miles or had the unit off for several months you will need to reinitialize it before you can use it.
Fails to highlight any satellites	If the above fix work then it may be that the unit simply won't receive signals and may need to be returned to Garmin for repair. It might be possible to perform a hardware reset, which may get the unit running again. See the chapter on secret commands for more details.
Fails to highlight any satellites on a multiplex receiver.	Occasionally these units will simply fail to find any satellites when first turned on. The solution is to turn off the unit and turn it back on. The second time it always seems to work
Unit fails to turn on	Check the batteries. Be sure they are installed correctly. It is also possible that you have water damage. Dry the unit out and try again. If this fails you may need to return the unit to Garmin for repair. If desperate you may want to read the chapter on secret commands and see if you can revive the unit yourself but you are likely to lose all user data.
Poor GPS Message or GPS signal weak is almost constant	Your location is marginal. Turn off battery save mode. You may need to find a better sky view.
Clock is off by more than 3 seconds	You do not have a current almanac in the unit, therefore you are not correcting for leap seconds. You will need to collect a new almanac by leaving your unit on with a clear sky view for 15 minutes.
Clock is off by more than a minute.	You don't have a lock yet or if you do the time zone setting is incorrect. Make sure you have a lock and if the problem persists check the time zone setting.

Symptom	Trouble Shooting tips
Unit turns itself off for no apparent reason on my motorcycle, off-road vehicle or bicycle	This can be caused by battery vibration. Some batteries are shorter and narrower than others. You can try a different brand of batteries. You can tape two together or place some tape around the batteries that will be on the outside to minimize vibration. You can gently pull the springs in the battery housing to make the connection tighter. You can ask Garmin for some small springs that will set between the batteries to make contact better. You can use external power. For 2 battery units you may add some foam to hold the batteries in place.
Max Speed shows really large speed	While the max-speed indicator is a reliable indication of the minimum "max speed" you have achieved since last reset it is unreliable when it comes to displaying the real maximum speed. This means you can use it to prove you have gone no faster than what it displays but you can't use it to establish a new speed record. As the satellites drift in and out of sight or you encounter some multipath interference your unit will compute a slightly different solution based on the available satellites and this difference can result in an instantaneous change in position. This change can confuse the max speed indicator into thinking you traveled rapidly to the new location. Luckily these are usually very easy to spot. Hopefully Garmin will figure out a way to filter out these kinds of changes in a future release.
Interface problems	Please see the interface chapter for hints on these kinds of problems.

Chapter 20
Selecting a New GPS

While most readers will have already selected their Garmin unit there may be some readers who want to study the book before buying one. If you fall into the later category then this is the chapter for you.

Choosing the Right GPS Receiver

One of the most often asked questions asked by a new prospective GPS user is: "What unit would you recommend and why?" It is impossible to answer this question without more information. There are many fine units on the market and what one person would like or need may be very different from what another person might need or want. This chapter will teach you how to make a good evaluation so that you can select a good unit the first time you try.

The Basics

A GPS receiver is a unit that computes your location using triangulation techniques. It gets information from 3 or more satellites and computes a position. As part of that solution it can also derive speed and direction data. It can save a certain amount of the data that it collects and/or computes for future use. Finally it offers a display that presents this data to the user. (Some units use a pc or a palm as the display device.) The computation is based on a set of moving satellites that circle the world twice a day so it can be used anywhere on the planet and up to 60,000 feet above the planet. But just a thin film of water blocks all signals so it can't be used under the water. The method of display and the kind of data it saves differ from unit to unit and is one of the things that separates units.

The user display often mimics devices that the user may already be familiar with which could fool a new user into thinking those devices are in the unit. For example a speedometer display matching the one on your car or a compass display that looks like a standard compass. Neither device is actually in a GPS receiver but are computed based on your current position and Doppler data available as part of that solution. So if you are moving the unit can look like it has these devices. It will also compute an altitude, which causes some folks to wonder if there is an altimeter inside. Again this is a computed output and the unit does not have a built in altimeter. By the way, due to the satellite position the altimeter reading isn't very good and can be up to 30 meters off and swinging wildly. Both a compass and an altimeter are good backup devices for a GPS when hiking.

Even so, there is no reason why a unit couldn't have these extra sensors inside and indeed the Silva GPS and some Garmin models do have a compass built in (and even an alitmeter) as does some survey grade receivers but in general inexpensive GPS devices do not. A GPS usually does have a built-in clock to help with its initial startup but after a position is computed the display of the time is based on data that comes from the satellites and not from the built-in clock. Using the satellite data for a clock provides you with a very accurate clock. Note that the behavior of these extra built-ins is different from the simulated displays in a standard GPS. For example the altimeter does not need a fix to work but may be calibrated using a GPS fix later. The compass can be used while the unit is stationary, unlike the simulated compass display, and will even know which direction you are holding the unit in while the computed compass knows the direction the entire unit is moving in. Often these are entirely different things.

The Features

At one time all GPS units attempted to cater to everyone. As units have progressed and needs become more focused manufactures have begun to produce units that are specific to an application. While there are still universal units in the market place there are more and more specialized units available. For this reason you really need to understand exactly what you intend to do with a GPS before you buy one. By the same token once you have a unit you will probably find uses for it that you didn't consider earlier so you will want a few features that are beyond the basic things you can currently think of. You may want to read the Introduction Chapter for some background.

A GPS receiver represents a quantum leap in technology over most items that you may have purchased. You need to spend some time understanding the technology behind the GPS system in order to determine if it will do what you need and to help you understand what to look for in your purchase. Unless price is really an important consideration go for a 12 channel parallel unit. (See the next section for why.) If you want the receiver to work without waiting around for 2 or more minutes with a clear view of the sky or to use it in applications where there is not always a clear view of the sky then get a 12 channel unit. A unit that can work standalone is usually a better choice for most users unless your application always needs to have a computer along. A standalone unit can work with a computer but a unit without a display won't work without one.

Here are a few things to help with your evaluation: The best thing you can do is to spend some time working with the proposed unit in your application. Or if that is impossible the see if you can try it out in the store. Familiarize yourself with the interface and decide which is more comfortable for you.

The questions

Questions that you may want to ask.

- ❏ Does it have the specs you need? Don't go overboard in weighing this. Most GPS receivers have very similar specifications.
- ❏ Can you easily find the data you need?
- ❏ Do you have to change pages more often than you like?
- ❏ Does the data have all of the formats you are interested in (grid systems, datums). If you want it to work with paper maps then learn about this before buying a unit.
- ❏ How's the configurability, customization?
- ❏ If you want a computer interface then does it support one? (All Garmin's do)
- ❏ Can you get the programs you want to work the GPS?
- ❏ How is the screen display? Is the text easy to read?
- ❏ Do you need an external antenna? Most of the time you don't but you need to evaluate this for yourself.

❑ Do you want or need internal maps? Do you need to update them or provide more detail?

❑ Is a waterproof receiver important?

❑ User interface ok?

❑ How's the battery life?

❑ Available accessories, mounts, cases?

❑ If you need built-in maps then how accurate are they? No current map databases are as accurate as you probably thought they were.

❑ Can the unit be upgraded? In this high tech market firmware upgrades are important.

❑ Do I need DGPS? - an optional hardware unit that provides increased accuracy.

❑ Do I need WAAS?

To get answers to these questions you can ask the salesman, but you may not be able to trust the answer unless you get one that understands GPS receivers. A better route is to go to the manufacturer's web site and look up the data. They often have manuals available on-line so you can download and read about the unit you are considering. If you have friends with GPS receivers that talk to them but learn a little first so you can speak the same language. Many GPS users claim that the unit they decided on is the best. This may be true, but filter out the reasons and make sure that it is the best for what you need.

There are also some questions that you don't really need to ask:

• Does it work outside the USA? All receivers work anywhere in the world except units that only work with a laptop using a map display. Some of these won't work outside the USA only because they need a map and may not provide any non-USA maps. This is another good reason to select a standalone unit.

Your applications

Be sure and align your expectations with what a GPS receiver can really deliver. It cannot reliably deposit you at your front door but can locate your yard. It will not replace a compass in all applications but can if you are driving in your car. It could replace your speedometer on your car but may not be as accurate at low speeds like hiking or on your sailboat. Newer units do better in this regard.

Typical applications for GPS include automotive, RVing, hiking, biking, aviation, motorcycles, and marine. Other categories include skiing, snowmobiling, and off-road driving. You need to decide how many of these you intend to use the unit for and what is really important. Many units have been specifically targeted at customers in specific categories. If you have specific mounting requirements then check into this as well.

A navigation receiver is not a surveying device but it is likely to be more accurate than most of the digital maps you use with it anyway. For more accuracy you can add a separate beacon receiver to supply correction data to the GPS (some receivers have a beacon receiver built in but it still needs a separate antenna) or you might buy a unit that has WAAS capability. WAAS can provide corrections similar to a beacon receiver and uses the same antenna as your GPS receiver. It, however, is line of sight and does not offer as good a reception as a beacon receiver in areas covered by beacon receiver. In addition WAAS depends on geo stationary satellites which may be fairly low in the sky depending on your latitude so they may be blocked by buildings, trees, or hills. Both solutions can get your accuracy down to 3 meters or so, which is still not survey accuracy.

Most of all, enjoy your search. Welcome to the world of GPS.

Why buy a 12 Channel GPS Receiver?

Many people wonder which GPS to buy. Some very basic units have a single or dual channel scanning receiver that can scan and 'lock' on 8 to 12 satellites. A frequently asked question is "Why buy a 12 channel GPS receiver?

This section answers the questions regarding 12 channel receivers and why you might want one. Note that sometimes this discussion will refer to satellites as SVs (Space Vehicles) and when the term GPS is used it will usually mean the receiving unit and not the entire system.

What is a 12 Channel receiver?

This is the first question you need to ask since there are two different interpretations of the answer. One is, any receiver capable of tracking 12 satellites. The second answer is a receiver having dedicated hardware to receive 12 channels **simultaneously,** also known as a twelve channel parallel receiver.

What kind of receivers are there?

Today there are basically three kinds of receivers. Multiplexing receivers, sequential receivers, and parallel receivers. There are also mixed types where some channels are dedicated to a single satellite and some channels are multiplexing or sequential.

What is a parallel receiver?

The receiver has dedicated separate hardware to receive each satellite that it needs for a solution.

What is a multiplexing receiver?

This kind of receiver uses a single, dual receiver, or even 3 receiver hardware design to time division multiplex among the satellites that it is viewing. This mean it gathers some data from one SV for a slice of time and then switches to another SV to gather some more data. If it is able to perform this switch fast enough it seems to be tracking all of the satellites simultaneously.

What is a sequential receiver?

This kind of receiver also switches limited receiver hardware among all of the satellites. It, however, gathers all of the data from one SV before moving on to the next as opposed to the multiplexing unit that switches using a time slice algorithm. Sequential units tend to offer sluggish performance, particularly with the first fix, since, if it has some trouble getting information from one SV, it can get bogged down before giving up and switching to the next SV. This unit can give good performance if it has a minimum of 3 channels. You are unlikely to find a new sequential receiver. Garmin never made one of these.

Is a parallel receiver better than one that time division multiplexes?

Parallel receivers are faster and will generally provide a more reliable fix. However, multiplexing receivers are fine under many conditions where there is a clear view of the sky. Under marginal conditions, such as tree cover or in city canyons, having a dedicated hardware channel remaining continuously synced to a particular satellite can be advantageous.

If I want a parallel receiver, how many channels do I need?

The minimum number of parallel channels you need is 4 to obtain a fast fix since a 3D solution needs 4 satellites. But that doesn't leave any spares to look for other SV's that may be just coming into view to replace the ones you are using that are about to drift out of view so you really need a minimum of 5. There have been 5 channel parallel receivers made and they work pretty well. Hardware costs have changed the picture and it is now easier to just dedicate one channel to one SV than to try and multiplex receivers around among the available SV's.

Why do I want more than 5 parallel channels

Since 5 would only allow for a margin of 1 SV having more channels would permit you to maintain a solution when you moved behind a building where you suddenly lost several SV's. Ideally you would like to track all available satellites simultaneously to maintain the likelihood of a fix under the worse possible conditions. Thus, if you turn a corner and your automobile obscures several SV's there is a good likelihood that the extra channels will instantly bring several others "on line" thus maintaining continuous lock.

Are there any other reasons to track more than 4 at once?

Given that you want all parallel receivers then you get an added benefit of being able to calculate an over determined solution (not all will brands do this but all Garmin 12 channel units do). An over determined solution uses extra satellite measurement data to help improve the accuracy of the solution.

Why aren't all units parallel?

Cost. Back when hardware was more expensive, multiplexing and sequential receivers were widely used and could track 6, or 8, or even 12 SV's by switching among them using GPS units equipped with from 1 to 5 receivers. The performance was OK but not nearly as stable and reliable as the 12 channel dedicated receivers. Today, the technology has advanced to the point that almost all receivers are parallel designs.

So why 12 channels?

The government maintains a minimum of 21 working satellites in orbit around the world. To ensure 21 working satellites there are usually 3 working spares. Sometimes there are more, currently there are 27 working SV's. With 24 in the sky there is some likelihood that 12 might be in your hemisphere at once. In practice many people have reported that, under some conditions, they have tracked 12 simultaneously for short periods. Often you cannot track nearly that many, sometimes as few as 6 SV's may be present in the sky.

In addition some satellites may be in view but are so low in the sky that their signal is not reliable and couldn't be used in a solution anyway. Therefore 12 is a good maximum number but isn't a gating item for performance. The fact that most 12 channel units are also parallel receivers is what makes them great choices.

Are 12 channels good for multiplexing receivers too?

Since multiplexing receiver do not compute an over determined solution the tracking of 12 channels is less important for these units. Generally a multiplexing receiver picks the best 4 SV's and computes its solution. Tracking 4 more means that if all of the originals were lost for some reason there would be a full replacement available. Tracking additional SV's can actually eat up some of the time slices and make the unit less responsive. Each manufacture designs their unit to maximize the performance or marketing potential they can deliver. Most folks find no difference in a unit that tracks 8 versus a unit that tracks 12.

How do I know what I have?

Check the documentation for your unit. Sometimes however the manufacturer doesn't tell you. The second method of determining whether you have a parallel unit or a multiplexing unit is to determine the time for the first fix of the day. To get the first fix the GPS must download ephemeris data from 3 SV's to compute a 2D fix (4 for a 3D fix). The needed ephemeris data is transmitted every 30 seconds from each satellite in orbit. A multiplexing receiver will need 90 seconds (3 times 30) to obtain this data and then some time to compute the fix. A parallel receiver can receive the data in parallel and will thus have all the data for a 2D or 3D fix in the same 30 seconds. The specification for a multiplexing receiver is usually about 2 minutes for the first fix of the day while a parallel unit can get a fix in about 45 seconds. Thus it is pretty easy to figure out what kind of unit you have.

Aren't 12 channel parallel units more accurate?

Not necessarily. The accuracy specifications for both parallel and multiplexing units are often the same. In practice a unit that can calculate an overdetermined solution is likely to produce a slightly more accurate position.

Garmin Choices

With the large number of models available from Garmin there can be real confusion as to which one to select. This manual has divided the available products into families and this grouping can have some merit in product selection as well.

The 12 family is a group of older 12 channel units. They are excellent units and likely the most popular that Garmin has ever produced. They make sense for users not needing a mapping receiver. These are reliable workhorse products but have largely been replaced with newer designs.

The III family was Garmin's first introduction of mapping receivers. These proved very popular with users and the III+ added user uploaded maps which really set the stage for a whole line of mapping products. The ability to rotate the screen display makes these units a good choice for users wanting vertical orientation for hiking and horizontal orientation for vehicle use. The II+ offers this feature in a non-mapping unit. The main limitation of the III+ is the limited amount of memory available for map upload (1.4M). Today the GPS-V takes this case design and upgrades it to a state of the art receiver with 19 Meg of memory and autorouting on the unit itself. This one feature may dictate that this is the receiver of choice for many people.

The emap is aimed directly at the vehicle use market and really focuses on map usage. It does not have an autorouter on-board but does have unlimited map storage via pluggable cartridges. The pluggable cartridge is the primary reason that this unit is not as waterproof as some of the other models but folks still like its size and shape for hiking use. Garmin developed a neoprene sleeve to go over the unit for these users.

The etrex line is intended for hikers and other outdoor enthusiasts. It is a very small and lightweight unit with specialized features for this market. The screen size is too small for use in a vehicle except by the passenger. The size can be a slight negative under heavy tree cover since the antenna is a bit smaller than the ones in the other models.

The 76 line is the most universal of all the modern units and is the favored unit for marine use as well as users who want one GPS for a wide variety of tasks. These units support an optional external antenna, which can make them the favored unit even for hiking in very dense woods. The new 72 could be considered to be the baby in this line of receivers for users who do not need an external antenna.

The first decision the user needs to make is whether maps on the receiver are important to them. To some users a GPS receiver means a moving map display so that makes the choice simple for them. On the flip side the screen size is pretty small and mapping receivers are more expensive. In addition you still have to buy the maps unless only very basic mapping is needed. Some prefer a laptop to display electronic maps or tried and true paper maps for use with their GPS.

All of the Garmin handheld receivers sold by Garmin today are 12 channel parallel units, but if you buy a used unit you could still find a multiplex unit.

Chapter 21
Other Garmin Receiver Models

This manual has been directed at only a subset of the GPS models available from Garmin. This chapter will try and fill in the gap by describing a few other models and how this manual may be used to help in their use. Garmin has made many models over the years and even in this chapter it would not be possible to cover them all. This chapter covers a subset that includes mainly current models and only models with some connection to other units that are covered in this manual.

Garmin has segmented the GPS market into 4 pieces, Aircraft, Marine, Outdoor and Mobile. Since this manual is targeted at handheld units it covers some units from each of the above classifications but generally the smaller units in each field. The Outdoor market is almost fully covered since without a vehicle to hold the unit they must be handheld units. Garmin also makes a few units that are merely GPS engines and have no display capability at all.

Mobile

The main units in this category include the Street Pilot, the Color Street Pilot, and the Street Pilot III. They are most closely related to the emap handheld and information aimed at that product in this manual will often apply to these units. The Street Pilot III is similar to the GPS V product.

The Street Pilot has a large 160x240 pixel horizontal gray scale display. The unit is a mapping receiver and supplements a basemap with data cartridges that are interchangeable with the emap. The **Color Street Pilot** is contained in similar cabinet to the street pilot and has the same style user interface. The screen is slightly smaller but boasts a full 16-color display. This was the top of the line in Garmin mobile units when it was released but has been largely superceded by the Street Pilot III.

The user interface on these units has many more keys that on a handheld unit. This permits dedicated function keys. These keys include:

Power key - which doubles as the lamp/contrast key like on the 12CX.

Page key - standard functions

Quit key - standard functions

Enter key - standard functions

In and Out keys - zoom

A rocker keypad

Find key - same as the emap.

Option - StreetPilot name for the Menu key

Route - A unique key on this product to create and edit routes and start/stop navigation.

Mark - standard function

The screen display has a very limited number of pages. There is a map page, a trip computer page, which includes speed, and if navigating a route page (called a road sign page). The trip computer page can be toggled to show satellite status. Most commands work like the emap or like the III family however the Street Pilot is missing a backtrack capability. The routes can use waypoints, which can have up to 10 character names, or mappoints like the emap. This is important since there are only 100 user defined waypoints. Once a route is enabled you get turn by turn instructions from the road sign page and other places. While the Street Pilot won't do automatic routing it will find a location by address or intersection or poi with the right MetroGuide maps installed and you can easily create a route just by rubberbanding the line drawn from where you are to where you want to go.

These units will run off of 12-volt power or 6 AA batteries. The Street Pilot will get up to 16 hours on a set of batteries while the color unit gets about 9 hours if you keep the screen as dim as possible or about 2.5 hours on maximum brightness. Color takes a lot of power. They all use memory cartridges for map memory (except the basemap) which are changeable by the user. Memory cartridges range from 8 Megs to 128 Meg however the larger sizes (over about 32

Meg) are really only useful for the Street Pilot III since the earlier units have a limit of 50 maps.

The **Street Pilot III** paved the way for a new standard in portable units. It features autorouting right on the unit itself and has a 305x160 pixel color display that is 3.4" wide and 1.8" tall. It was followed by its smaller handheld brother the GPS V which also has autorouting on the unit. For a discussion of autorouting see the autoroute chapter, which describes this feature from the GPS V implementation. The Street Pilot III takes the best features of the Color Pilot and adds a more powerful processor to be able to provide routing directly on the unit. Like the Color Street Pilot it uses heuristics to customize its arrival time estimates when running a route based on the driving habits of the owner. Unlike any other unit in the Garmin line it offers turn by turn instructions with a human voice. It also has a capability of storing up to 500 waypoints and features a 2000 entry tracklog which makes it comparable to handhelds.

Other Handhelds

Garmin makes two units that they call handhelds that have a very large display. These are the **GMap175 marine unit** and the **GMAP195 aviation unit**. They weigh in at a whopping 1.4 lbs., like the street pilots, and use 6 AA batteries with an expected battery life of about 10 hours. They also support a rechargeable battery pack. While they can be held in your hand you wouldn't want to carry them very far. They use a display like the one in the Street Pilot but oriented vertically. Both are mapping receivers and do have a basemap but for optional maps they only use pre-loaded cartridges available from Garmin. The cartridges are specific to the intended use of these products. Marine cartridges are available for the gmap175 that provide charts covering most waterways all over the world. These G-charts provide depth, shoreline, and navaid data. The gmap195 uses aviation maps covering distinct areas and has full Jeppesen data plus Final Approach Waypoints for most airports. These units are most similar to the III family of handheld units except they have extra dedicated keys. The 175 includes a MOB key and a Mark key in addition to those of the III family while the 195 uses those keys for NRST and WPT. These keys access the nearest and full databases including all of the Jeppesen data and user waypoints. Press the WPT key twice to actually set a waypoint. There are only 256 user waypoints available for these units but this is augmented with an extensive set of navaids.

The 175 has a full featured Highway display for navigation while the 195 features an HSI navigation screen. These screens are similar to those described in the navigation chapter except that the 175 Highway display includes all of the programming ability of a III series unit and all the capability of a G-45XL unit including CDI plus an edgewise compass display. Both internal and external alarm capabilities are supported.

Garmin also makes a smaller aviation unit called the **G-92**. It is physically very much like the G-48 unit and has similar specifications to all of the units in the G-12 family, except that it includes the full Jeppesen database instead of a city database. It does not provide a basemap but does provide a graphic display of airspace.

Marine Units

Besides the handheld units covered in this manual Garmin makes a line of units that are designed to be permanently mounted in a boat. They include units that are very similar in operation to the G-II+ except the display and keyboard is larger. They may or may not be combined with a fish finder. Two of the latest offerings are mapping receivers called the **MAP162** and the **MAP168 sounder**. These units have large displays and contain basemaps similar to the ones in the III+. They have 2.5 Meg of map memory and can use the same Mapsource maps that are available for Garmin handhelds. The 168 is similar to the 162 but includes a built in depth finder that is integrated with GPS operation. Both units support depth data rather than altitude as might be expected in a marine unit. They will even provide celestial data on the sun and moon and calculate the tides.

Aviation Units

Garmin makes a line of panel mount units that are dedicated aviation models. These units are unlike any of the units talked about in this manual except that they share a similar technology. In addition they make another portable unit called the GMAP295. The **GMAP295 aviation unit** is a top of the line portable GPS receiver. It is at home in a plane and provides features that make it also useful in a car or other mobile application. The case style is similar to the Street Pilot and it has the same Color display as the Color Street Pilot. It has a full basemap and contains 2.5 Meg of memory for map loads using Mapsource maps. In addition it has the ability to use memory cartridges like those in the emap and can therefore support MetroGuide maps for street use. It supports all of the features of the III Pilot that is covered in this manual and can even split the screen to display two different pages simultaneously.

TracPak

Garmin makes a line of small units that do not have a display. They can be used with a laptop or pda supplying the user interface and require external power. They can be purchased as a card or as a small mouse like box or even in a case that sets atop a pole or mast, the GPS36. They share the engine technology with the other Garmin units but must be controlled only via the user port interface. Some of these units have two serial ports. The GPS35 series is a family of units that can be used in a variety of applications. Most units are intended to be hardwired or provided with a user supplied connector. The **GPS35PC** is is a mouse like unit and is designed to attach directly to the rs232 port of a personal computer and includes a cable for an external 12-volt power source. The latest mouse like GPS unit is the **GPS16,** which includes WAAS capability. It has an RJ-45 interface connector. OEM suppliers sometimes use TracPak units. There is one pcmcia device that uses one of these units.

Communications plus GPS

Garmin makes two unique navigation receivers called **NavTalk** that are bundled into cell phones. There is one model for land use and one for aviation use. These units provide most of the functionality of a III+ receiver with the added functionality of an AMPS compatible (analog) wireless phone. In addition to being able to use the two units separately the GPS position can also be sent automatically or by command from the unit. This position data can be received by a second NavTalk unit and displayed on its maps or can be available through some web based services that can provide this data on a moving map display.

One advantage of a combined unit is the addition of the cell phone keypad. The GPS portion of the receiver can take advantage of these extra keys to simplify data entry.

Garmin also makes a marine communications / GPS receiver that works on the marine VHF band called the GPSCOM 170 and a couple of FRS (Family radio Service) / GMRS (General Radio Service) units called **Rino**. The Rino 110 is a non-mapping unit similar to the etrex Venture and the 120 is a mapping unit similar to the Legend. The etrex discussion in this manual should be adequate for using this unit. The Rino name is an acronym for Radio Integrated Navigation Outdoors but the appearance of the unit with dual antennas (a quad-helix for the GPS paired with the UHF antenna for the radio) is clearly where the name comes from. The user interface matches the etrex models including the "click stick." For Canada the GMRS feature is disabled.

About the Author:

Dale DePriest first became interested in GPS in 1996. He soon acquired a Garmin 38 and began a love affair with this technology that continues to this day. Early in his thirst for knowledge he found a news group titled sci.geo.satellite-nav where knowledgeable folks discussed this topic and the units that could perform the magic of GPS. Since he had an engineering mind and background he felt he really needed to understand how these devices worked. A bit of reading and a lot of experimentation soon started revealing the secrets locked up in these handheld units. He started sharing the data he had found and providing some help to other users. This led to his connection with Joe Mehaffey via his web site and Dale's contributions became articles on the web site joe.mehaffey.com (now called gpsinformation.net). New purchases of a Garmin12, later an emap and a vista, provided connections to the latest 12-channel technology. Receivers such as these really moved the GPS into the main stream as a reliable navigation tool once you understood how it worked. As more users adopted this technology, Dale found that many questions were being asked over and over again which led to an idea of developing a manual on how a GPS worked and more specifically how to use one. He started by adding articles to his and Joe's web sites and, through the encouragement and feedback of readers, a manual started to form. Now, 4 years later, the baby has grown up and ready for the world. Dale continues to expand his knowledge and provide reviews on GPS related software and hardware. Combining his GPS passion with a new passion for handheld devices, he has built web sites for both palm users and pocketpc users, as well and continuing coverage of GPS topics. His site can be reached at http://users.cwnet.com/dalede.

Printed in the United States
23874LVS00003B/176